Intelligent Systems Reference Library

Volume 143

Series Editors

J. Kacprzyk, Warsaw, Poland
L. C. Jain, Adelaide, Australia

For further volumes:
http://www.springer.com/series/8578

Zekâi Şen

Philosophical, Logical and Scientific Perspectives in Engineering

 Springer

Zekâi Şen
Civil Engineering
Istanbul Technical University
Istanbul
Turkey

ISSN 1868-4394   ISSN 1868-4408   (electronic)
ISBN 978-3-319-34869-8   ISBN 978-3-319-01742-6   (eBook)
DOI 10.1007/978-3-319-01742-6
Springer Cham Heidelberg New York Dordrecht London

Printed on acid-free paper

Springer is part of Springer Science+Business Media (www.springer.com)

I would like to dedicate this work to all my students and future generations in different cultures at the service of humanity through my grandchildren:
Dilara Nur, Elif Derya, Hatice Ceyda and Mehmet Yigit

Preface

Engineering career is defined in many ways but all definitions have a common point that engineering practice benefits from the end product of scientific research, which excludes the artistic facets and confines engineering into a form where philosophical, logical, and scientific dynamics do not play significant role. Generative engineering innovations, ideas, and descriptions leading to intelligent design and planning should include science philosophical and logical ingredients which are like torches that illuminate toward better conceptions, hypothesis, artistic imaginations, and their implementations. In the past, for planning, design, construction, maintenance, and operation of engineering structures, engineers had innovative practical and creative artistic abilities by means of philosophy, aesthetics, and logical inferences even linguistically without any mathematical equations. Later, systematic education and training programs led engineers to get away from such abilities with more emphasis on analytical and numerical methods leading to regular and classical deductive inferences as standard solutions. Although analytical intelligence and ability are indispensable ingredients for mass production, their crisp and hard rule applications do not provide creative bases in engineering career. Engineering career can be considered as a group of civilization establishing specialists who deal with a set of interconnected road network, bridges, park, infrastructure, subway, construction, space shuttle, off shore oil platform designs and their hardware features based on convenient scientific, mathematical, economic, social affair, and practical knowledge availabilities. During the evolvement of these activities conceptualization, imaginative design, and idea generations should take joint shares, which can be achieved only through inclusive philosophical, logical, and finally scientific principles considerations.

As for the present day engineering education one can easily say that philosophy of engineering is almost nonexistent in engineering curricula. Another vague entrance of philosophical thinking into engineering domain may be that engineering and technology are thought as distinct disciplines, but there are occasional interferences between the two, which transfers some philosophical aspects into engineering thinking, because some engineers and numerically trained technicians took active involvement in technological developments such as in the first atom bomb production. Some may propose that there is no need for separate philosophy principles in engineering, because they are included in the philosophy of technology. This is not acceptable due to the benefit of engineers from the end products of science rather than technology in their generative mental outputs such as design, solutions to many problems, and model constructions. Most of the engineering students in the world cannot write proper reports or articles due to their inefficient philosophical and logical backgrounds, because they are empowered mostly by numerical, mathematical, and crisp logical foundations.

Engineering career has penetrated into a variety of human activities that it become necessary to look at its very roots for creative reasoning leading to rational and productive products. It has already been noticed since the last decade that this career is away from the basic philosophical reasoning that should be inferred through the logical principles not in the form of final symbolic mathematical equations only that help for ready engineering calculations, but initially in terms of linguistically propositions. Although the end products of scientific researches in the forms of equations, formulations, algorithms, and software are essential products for engineering career, without their logical and verbal expressions, it is not possible to communicate let alone among the same engineering disciplines, but even among the experts, who are capable to grasp linguistic explanations. In order to eliminate such rigid situations and provide a common linguistic discussion domain among all types of expert views, recently science philosophy version of engineering philosophy and its subsequent activity, logic with its propositions started to take place among the engineers toward engineering philosophical aspects. For this purpose, advanced engineering principles based on the philosophical reasoning and logical inferences are bound to play significant role in future engineering activities. Such ingredients are necessary for innovative creative end productions with research and development activities.

It is the main theme of this book to emphasize the significance of philosophy, logic, and science in engineering education and training. Without these bases engineers become addicted to case study solutions, ready software or convenient formulation matching during problem solving stages. Most often creative abilities are not cared for future improvements and advancements. Philosophy of science provides dynamism for the creative intelligence of engineers. This book provides a forum for engineer's harmonious integration with engineering aspects toward effective model constructions that are capable to solve a spectrum of problems with different methodologies. The content of this book has been delivered as a course on postgraduate level under the title of "Engineering Research Methodology" at the Technical University of Istanbul, Faculty of Aeronautics and Astronautics as well as in Civil Engineering Faculties. I could not complete this work without the love, patience, support, and assistance of my wife Fatma Şen.

Istanbul, Erenköy, 2013   Zekâi Şen

Contents

Chapter 1
Introduction

1.1 General

Engineering services and structures have been and still they are among the corner stones in any civilization growth and development for the social harmony [24]. In the planning, design, construction, maintenance and operation of these structures, engineers had innovative practical and creative artistic abilities with more dependence on philosophy, aesthetics, and logic in the past, but systematic education and training programs led them to get away from such abilities with more emphasis on analytical and numerical methods leading to regular deductive inferences as standard solutions. Although analytical intelligence and ability are indispensable ingredients for mass production, their crisp and hard rule applications do not provide creative bases in engineering career. It is the main theme of this book to emphasize the significance of philosophy, logic and science in engineering education and training. Without these bases engineers become addicted to case study solutions, ready software or convenient formulation matching during problem solving stages. Most often creative abilities are not cared for future improvements and advancements. Philosophy of science provides dynamism for the creative intelligence of engineers.

In recent engineering curricula, philosophy and engineering seem in conflict, because science philosophy is driven away almost completely from engineering education. In the curricula, philosophical reflections are almost non-existence in engineering training. So, how can one expect engineers to incorporate in their post-graduation life generative works ornamented with artistic, technologic and humanitarian aspects? Even though the philosophy of science is now well established, it still has mutual exclusiveness from engineering thoughts. An engineer can have appreciation, willingness and even applications according to individualistic philosophical thoughts, but it is not a general tendency in engineering educational systems almost all over the world. Since 1980s ethics entered into the engineering curricula as a part of philosophy [5]. However, other branches of philosophy as ontology, metaphysics, epistemology and aesthetics, also play roles in many engineering aspects and thoughts.

Z. Şen, *Philosophical, Logical and Scientific Perspectives in Engineering*,
Intelligent Systems Reference Library 143, DOI: 10.1007/978-3-319-01742-6_1,
© Springer International Publishing Switzerland 2014

Philosophy and engineering aspects can be given a rational inference mode through logical principles. Although in engineering crisp (two-valued, Aristotle) and symbolic logics are in use, since the last four decades a new candidate of logic entered the engineering automation and modeling subjects under the fuzzy logic rules. Its influence is expected to increase gradually in future engineering applications by means of expert view usage in many technological innovative studies. Engineering is understood as a career, where science end products are used for practical problem solutions. This statement implies that engineering is not a scientific task, but science is at the service of engineering.

This book concentrates on philosophical, logical and scientific principles importance for engineering works. It provides a forum for their harmonious integration with engineering aspects towards effective model constructions that are capable to solve a spectrum of problems with different methodologies.

1.2   Engineering

Aleksander [1]gave a simple definition of engineering. Engineering and architectural thoughts are related to metaphysical and ontological issues such as imagination, description and visualization of a certain problem along the solution path, and therefore, philosophical and logical interpretations and inferences are necessary to reach at a final solution among many alternatives leading to an effective decision making process. Such philosophical ingredients in engineering are relatively rare compared to the methodology, epistemology, ethics and aesthetics. A common ground among the engineers and philosophers includes topics of ethics, aesthetics, epistemology, methodology and ontology. These two groups of specializations have shown common interest during few decades only on ethics and since many centuries on aesthetics, but not on other issues. Perhaps, the first impression in both specializations is towards looking onto each other as if the interests are different, but as long as linguistic issues are concerned, engineers must shift towards the philosophical domain so as to increase their ability to draw meaningful conclusions for their current problems prior to numerical solutions. Coupled with the preassembled linguistic (philosophical and logical) understanding, any formulation or equation appears as a matter of dynamic activity on engineer's behalf.

Engineering career can be considered as a group of civilization establishing specialists who deal with a set of interconnected road network, bridges, park, infrastructure, subway, construction, space shuttle, off shore oil platform designs and their hardware features based on convenient scientific, mathematical, economic, social affair, and practical knowledge availabilities. During the evolvement of these activities conceptualization, imaginative design and idea generations should take joint shares. The needs of a society within a civilization are planning of urban area safety operation and management by realization of possible, rational and plausible alternatives. Many engineering colleges in the 1990s are busily revising the style and substance of engineering curricula to provide increased attention to design. The intent is to redress what many reformers see as an imbalance caused by too much emphasis on the analytical approaches of engineering science [22].

The American Engineers' Council for Professional Development (ECPD, the predecessor of ABET) has defined engineering as "the creative application of scientific principles to design or develop structures, machines, apparatus, or manufacturing processes, or works utilizing them singly or in combination; or to construct or operate the same with full cognizance of their design; or to forecast their behavior under specific operating conditions; all as respects an intended function, economics of operation and safety to life and property [6]. Engineers should have been trained for problem solving using their classical and educational skills in addition to knowledge that should be filtered through the following steps.

- Verbal and linguistic thinking (philosophy);
- Rational reasoning (logic);
- Application of methodological principles (science);
- Benefit from mathematical formulations (model);
- Use of approximate reasoning (expert view);
- Depend on professional experiences (practical view).

In the implementation of these steps the final purpose is to search for the best, rapid, simple, economic and optimum solution. Engineers in the past, without any deterministic and systematic education of the present day, were weighting all these steps at different percentages in their search for final solution. Today engineering education has fallen into the traps of ready formulations and software almost without sufficient reasoning. It has become a mechanical task to solve problems according to standard principles, algorithms, methodologies, software and models. They have lost the flexibility given by the aforementioned points and continue to their work between crisp boundaries of science and mathematics that came out from scientific activities. In this manner, engineering became a branch of applied mathematics and direct applicants of scientific outputs. Engineers started to run after appropriate mathematical models for problem solutions, which allow only to analyze and to test potential solutions. If the convenient solution could not be obtained then they try to fiddle (train) the available methodology to fit the desired output by playing with model parameters. Standardization in engineering aspects leads to classical and stand stagnant concepts that make them to forget the significance of the aforementioned points where scientific principles are applied for the optimal conversion and conservation of natural resources into structures, machines, products, systems, and processes for the benefit of mankind. Similar definition is given also by the McGraw-Hill Encyclopedia of Science and Technology (1997). As Lewis [6] has stated:

> The upshot of philosophical attacks would be to replace this traditional self-understanding with one that might read more like the following: "Engineering is the scientific art by which a particular group of human beings destroys nature and pollutes the world in ways that are useless or harmful to human life".

Many criticize that "engineers cannot be scientists", which is true if s/he is unaware of philosophical thinking principles without critical views but dependent only on scientific and mathematical propositions. For an engineer to become a scientist s/he should be empowered with principles of philosophy and logical propositions leading to rational inferences. Philosophy and logic necessitate linguistic (verbal) means, which are available rather vaguely in engineering educational systems, which are full of numerical formulations, symbols, equations and software. The first intensive contact of engineers with logical principles is due to software writings, where even slight logical error causes mistakes and the software cannot produce desired and correct outputs. Debugging in any software development requires logical principles more than any numerical calculations.

Mitcham and Mankey [19] gave the general characteristics of engineering through on more linguistic side as follows.

> What engineering is might be better determined by how the word "engineering" and its cognates and associated terms (such as invention, innovation, design, technology, science, etc.) are used, especially in relation to each other. From a linguistic philosophical perspective, it would be appropriate to begin not so much with our experiences of engineering, but with the words we use to talk about such experiences.

On the other hand, Davis [5] is against the idea of philosophical definitions of engineering and a linguistic approach. He suggest engineering definition as,

> ... all attempts at philosophical definition will, (a) be circular (that is, use "engineering" or a synonym or equally troublesome term): (b) be open to serious counter-examples (whatever because they exclude from engineering activities clearly belonging or because they include activities clearly not belonging: (c) be too abstract to be informative: or (d) suffer a combination of these errors.

After such criticisms he suggests engineering definition from the historical point of view as follows.

> Engineering, like other professions, is self-defining (in something other than the classical sense of definition). There is a core, more or less fixed by history at any given time, which determines what is engineering and what is not. This historical core, a set of living practitioners who—by discipline, occupation, and profession—undoubtedly are engineers, constitute the professions.

Philosophical definition of engineering cannot be acceptable by its own, but engineering activities will be more productive if philosophical aspects are intermingled into these activities. Engineering can be defined more as an art and ability of rendering available natural resources for the service of human after combined effects of philosophy, logic and scientific inferences. In this statement, there are two words as "ability" and "art", which can be explained only philosophically. Scientific theories and results are produced by scientists; as end product user engineers cannot be scientists, but by acquainting themselves through philosophy of science principles, they can came into close contact with scientists. After all, scientists who are empowered with the philosophy of science and any engineer with background on the philosophy of engineering can find a common hinterland for discussion. Unfortunately, without philosophy of engineering, engineers are confined to restrictive applications of scientific outputs only. In order to avoid such a situation, engineers must be empowered with philosophical thoughts in addition to ethical and aesthetical implementations. Focusing on philosophy of engineering together with analysis of technical functions is presented by Vermaas [25].

Classical engineering solutions do not consider the philosophical principles, which help to explain wholeness. One of the main concerns in engineering is the economic consequences. Such solutions are not considered within wholeness, and therefore, may carry harmful ingredients. Instead of trying to improve the occurrence of harmful cases afterwards, it is far better to search for solutions by considering philosophy of engineering principles right at the beginning from different facets, and hence, to suggest engineering solutions based on linguistic information and subsequent logical rules. Mathematical formulations and models have logical and linguistic information at their bases, and engineer must put forward his/her thoughts linguistically into action with his/her set of knowledge about the subject concerned. Philosophical knowledge provides reasonable distinction between the useful and harmful aspects of a problem, where useful aspects are selected rationally. Today engineering curricula include socio-economic and cultural courses for linguistic training of engineers, but without philosophy of engineering such an approach may not reach its target successfully, because in engineering all the symbolic and numerical results can be useful only if they find their linguistic counterparts in engineer's mind and memory. Additionally, philosophy of engineering principles is bound to provide a more dynamic basis for better understanding of socio-economic and cultural courses. Luegenbiehl [13] defines engineering as:

> The transformation of the natural world, using scientific principles and mathematics, in order to achieve some desired practical end.

In this definition, the words "scientific" and "mathematics" as preliminary requirements for engineering implies "philosophy of science" and "logic", which are also disciplines under the philosophy. Unfortunately, many engineering institutions all over the world do not care for philosophy of science or logic. They drive away these two major legs of the modern engineering career and concentrate more on science products for the sake of science and mathematics. Consequently, engineers seek solutions to their problems by using the end products of the scientific achievements and mathematical end products in terms of formulations, equations or algorithms without linguistic bases, where philosophy can provide a creative thinking domain. They know superficially that mathematics means logical principles and science finds its bases in the philosophy.

In general, philosophy, science and engineering aim at geniuses, truthiness and practical ends. Search for practical ends do not stimulate philosophical reflections but rather practical rules of thinking only. Knowledge acquirement is an integrated part of engineering, which can be achieved through theoretical scientific researches. For this reason, many theories in physics, chemistry, biology, mechanics, etc., can be actuated by engineering laboratory setups only, which generate experimental data for further research in scientific deepness. Many scientific theories that seek knowledge about the world involve engineering activities in that endeavor; engineering should surely be of interest to philosophers and vice versa. There are pertinent questions to be asked about how the physical products of engineering can help to access knowledge about the world; what exactly is the role of manufactured objects in finding knowledge; and how reliable are they? These are questions that are dealt with by the eminent philosopher as Hacking [11] and also by few others. Engineering can be seen as delivering knowledge by much more direct route than by aiding science. There is a useful distinction in philosophy between 'knowing that' and 'knowing how'. One may know that Caspian Sea is in Asia, and s/he knows how to drive a car. This is an important distinction that is obscured by the single word 'knowledge', and when one takes it into consideration, it is clear that engineers seek to acquire knowledge in all of their endeavors.

Engineering is concerned with 'know-how'. Engineers know how to build a bridge that will carry traffic, and how to build a particle accelerator to carry out experiments. This latter kind of know-how represents knowledge relating to some of the most fundamental features of nature. Engineering as a consequence yields highly successful knowledge about how to control materials and processes to bring about desired results. It is a way of getting to the nature of things—a voyage of discovery as much as science. Hence, engineering provides a useful case study for philosophers inquiring about the status of human knowledge (http://www.stoa.org.uk/topics/engineering/philosophy-in-the-making.html).

1.3 Engineering and Philosophy

The first Philosophy and Engineering workshop is held in 2007 at the Delft Technical University, which is the first triggering work for the need for philosophical discussions in engineering. As engineering and technological developments pace for betterment, engineers need philosophy or philosophy and engineering start to meet each other in a more overlapping portion. In general, philosophers reflect ideas and draw out problems from problems, whereas engineers are more action oriented towards ending the problems in the best, cheap, simple, aesthetic and ethical manner. This implies that the engineers' involvement may also solve problems with new and further advancement problems in mind for the next application somewhere. Another difference between the two is that engineers are more objectively oriented through systematic algorithms, numerical models, computer software, graphical representations as plans, and equations, but the philosophers are language dependent seeking for logical inferences. The junction between the two then has language and logical statements in common; hence they can have a common platform only on the linguistic aspects. After all the engineering procedures have their foundations on the language, which unites the two disciplines on some parts of the philosophical principles. Unfortunately, today in many engineering research centers, universities and institutions, engineers are trained away from philosophical, i.e. linguistic rationalism, where there is an open door for the entrance of philosophical and then subsequently logical propositions. One can realize in any common meeting between engineers and philosophers that generally engineers make their presentations on electronic slides picturesquely whereas philosophers either talk naturally or from the paper, if some of them make electronic presentations most often they are in black-white slides without decoration. From philosophers' point of view any problematic issue can be settled down by a set of sound and sensible statements, but engineers require effective, efficient and sustainable problem solutions. After all what have been said as differences between philosophers and engineers, one may have the idea that they cannot come together on common issues, but the linguistic basis of all the information and knowledge prior to symbolic logic, equations and crisp decisions even engineers rely on language, and therefore, there are sub-common areas of interest between the two disciplines. So, engineering philosophy and logic should have harmonious combination from these disciplines to serve engineers more than philosophers or logicians. In a way philosophy of engineering is to try and identify philosophical aspects at the service of engineering, not vice versa. The author of this book believes that the entrance of even fragmental pieces of philosophy into engineering training will accelerate the production and innovation expectations from engineers. It is even not yet possible to establish rules and regulations about the philosophy and engineering coupled with logic in open literature, because such concepts are emerging recently in the scene.

Although science and comparatively to a lesser extent technology intermingle with philosophy, present day engineering has very limited overlap with philosophy (see Fig. 1.1). Engineering benefits from the outcomes of the scientific works, which help to develop technological innovations and scientific inventions, on the other hand, philosophy means marginal and almost non-existing contributions to engineering creative works.

Philosophy of science has an active role in the scientific studies since the science became independently spelled out from the philosophy during the renaissance period. In the last decades, through patent institutions, philosophy of technology started to ripen, and there are many articles in the literature about such aspects [8, 15, 21]. However, philosophy of engineering is a very recent debate in the world, since the last several years [4]. Hence, philosophy of science is an overlooked or delayed aspect for systematic engineering creative thinking. This

Fig. 1.1 Interaction among engineering, philosophy, science and technology

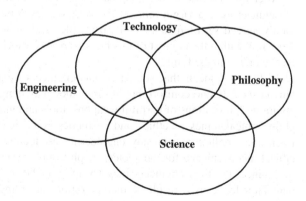

does not mean that engineers never benefited from the philosophical principles, of course they did, but it remained on an individual basis and it could not be systematized for many decades. For instance, ethical behaviors and career rules are based on philosophical bases, but real mass of engineering background on tangible aspects are far away from systematic philosophical principles. Even the entrance of ethics into engineering domain is a work achieved during the last few decades [5].

Figure 1.1 shows that engineering is a mixture of different disciplines and the best engineering training should adapt the relationship in different proportions between engineering, science, philosophy and technology [23]. Engineering depends on aforementioned disciplines but some of them are also engineering dependent such as technology and some confirmations of science can be reached after engineering experimentations. It is well recognized by many that science is concerned with discovery, technology with invention whereas engineering is more craft work concerned with making, producing and generating alternative solutions for a given problem. Critical scientific knowledge including theories falls within the domain of science, which is in continuous development throughout centuries. Patterns and blueprints are invention imprints in technological developments. However, engineering tasks are concerned with material products and designs.

One can easily say that philosophy of engineering is almost non-existent in engineering curricula. Another vague entrance of philosophical thinking into engineering domain may be that engineering and technology are thought as distinct disciplines, but there are occasional interferences between the two, which transfers some philosophical aspects into engineering thinking, because some engineers took active involvement in technological developments such as in the first atom bomb production. Some may propose that there is no need for separate philosophy principles in engineering, because they are included in the philosophy of technology. This is not acceptable due to the benefit of engineers from the end products of science rather than technology in their generative mental outputs such as design, solutions to many problems and model constructions. Most of the engineering students in the world cannot write proper reports or articles due to their inefficient philosophical backgrounds.

After all that has been explained above, one can understand that philosophy of engineering is a virgin topic at the verge of development and there is no commonly agreed set of fundamentals, principles and rules for the definition of this field. It is within the context of this book to propose such a set of items, which can be elaborated in the future.

It does not mean that scientists and engineers alike always try to improve human security and comfort, but at times they may dangerously destruct the society through inventions of harmful weapons, tools and materials. In such situations, philosophical topics as ethics and aesthetics come into view for interrogations. Even noise pollution from any engineering production can be considered as an ethical inconvenience for the society. If philosophical principles are not cared for in engineering, then engineers may try to save the day only through their stagnant knowledge as a result of memorization and blind applications. They may not heed for criticism, questioning, interrogation; they may apply their prescription on rather blind grounds. Any applied formulation, procedure or software is not criticized prior to application as for their suitability for the current problem. If the fundamentals of knowledge are not taken on philosophical and logical foundations, then the engineers may not even know how to achieve the work dynamically, which leaves domain only for dogmatic prescriptions.

Philosophy is a way of rational explanation of natural objects and phenomena, and an engineer, apart from the understanding of a scientist, must draw a share from this definition. Even though engineers are not scientists, they may come close to scientific findings by acquainting themselves with the philosophy of engineering. It is possible that an engineer can reach depths of scientific knowledge provided that s/he is interested in debates, criticism and interpretation of the end products from different philosophical and logical angles. Of course, an engineer should care for all these to end up with a successful application of available knowledge. One may have engineering background, but after graduation many years later, s/he may understand that engineering education, as it has been without philosophy of engineering and logical rules, cannot entitle an engineer to be a scientist. In various debates and discussions with many scientists without proper engineering background, one may enter into scientific discussions about what is the science? What are its principles? What are the features of scientists? After answers to each one of these questions, one may come to the point that "engineers are not scientists". If others realize one with engineering background and scientific qualities, then they point out exceptions. Such an exceptional stature is gained by trying to acquaint oneself with philosophy of science and linguistic logical rule generation. Extra interests give one ambition to continue to work on some interesting issues even after graduation, and finally, one also may came to the conclusion that engineers cannot be scientists until they have acquaintance with philosophy in general, and philosophy of science and philosophy of engineering in particular. Philosophy of science covers many aspects. Linguistic information, on the bases of philosophy and logic, can provide mathematical and engineering formulations' generation mechanism, but the reverse is not necessarily true and it is almost a dead end. Philosophy is not the property of any career; it has inter-career and service characteristics without distinction. In the classical education systems, philosophy of engineering principles is overlooked and engineering is considered within the restrictive confinement of economy, simplicity, speediness and practicality. Crisp engineering solutions may overlook environmental conditions and at the end harmful productions can take place without improvements after destruction, because the time may be too late. Presently, greenhouse effect, global warming and climate change are among such environmental phenomena.

Science and engineering can be distinguished because the former aims to build theories that are true, while the latter tries to make things work. The science is based on models or theories, but engineering deals with arte-facts or processes with scientific methodologies and methods. It is possible to understand the world through science and the change of the world is the main concern of engineers. On the other hand, the philosophy of engineering paves the antecedent conditions of both science and engineering. The main purpose of philosophy is to pursuit a question in order to find a genuine and knowledgeable answer. It is the task of a philosopher to search for such knowledge and then to check its confidence and reliability by logical principles. Genuineness and truthiness are among the aims of philosophy and science, respectively. It is the philosophers' duty to turn a critical eye towards science, to assess whether it really achieves knowledge, if so by what means, and whether one can ever be sure that the most successful scientific theories represent genuine knowledge about the world. In order to respond to some accusations and attacks on dogmatic engineering activities, engineers need to care for the philosophical foundations at an extent to defend themselves at least on ethical and aesthetical bases. On ethical and aesthetical bases philosophy becomes a crucial issue for engineers. The main topic of this book is not such a crucial affair only, but more significantly the importance of the philosophical issues within the engineering aspects of basic understanding concerning the scientific facades of training so as to bypass memorization and rather blind and rather dogmatic training subjects. Presently, the philosophical fundamentals are mostly unrecognized in engineering universities, institutions and colleges. Philosophy is important, because engineers faced with actual problems may not be able to solve them simply with crisp engineering methods alone. When confronted with problems that do not have spoon feeding solutions according to what they have acquired during traditional and philosophy immune trainings, engineers begin to realize their inability in the philosophical thinking. Most often engineers are confronted with questions such as in which way? and how? they should solve the problem that cannot be solved by technical knowledge alone.

Different subdivisions in the philosophy can be brought together by considering the following points and their mixtures to a certain extent [16]. He mentioned about the following steps as for the importance of engineering philosophy.

(1) Conceptual analysis: It is necessary for clarification and correction of terms in theoretical and practical uses, which involves basically logic,
(2) Reflective examination: It provides practice and thought, so as to deepen insight and understanding of extend, or to criticize both dimensions of experience. This includes the core areas of philosophy known as ethics, epistemology and metaphysics, often with an emphasis on their rational methodologies,
(3) Experience aspects: These are more global than customarily dealt with by any discipline. Such aspects may also involve inter-, multi-, trans-, and antidisciplinary consideration of what is right and good (ethics), knowledge (epistemology), and the structure of reality (metaphysics),
(4) The practice of a distinctive way of life and thought: It can be taken to be good in itself, with its own unique knowledge of reality. Philosophy in this sense may also be regionalized into the general guiding practices or principles of an individual or group, as when one refers to someone's personal philosophy or the philosophy of a firm.

Apart from classical ethical and aesthetical aspects of philosophy, in engineering, some other issues are also related to philosophy, but they are not very obvious or not used frequently by engineers [13]. Since philosophy is linguistic in its character, it is necessary to consider epistemological aspects of each terminology and question their logical contents in order to arrive at successful decisions. Epistemology is a sub-branch of philosophy and engineering is also related to philosophy through this branch.

Linguistic information and knowledge are capable to produce symbolic expressions that fall within the domain of mathematics. On philosophy side, one should pay attention to the reality of engineering practices, make debates on analytical and if possible especially on empirical grounds. Immediate result expectations by engineers do not give way to enter philosophical issues into engineering affairs, and therefore, most often without any questioning, interpretation or criticism engineering education empowers engineer candidates in such a manner that they seek ready crisp answers by using one of the methodologies through symbols in any equation or model. This is tantamount to saying that crisp thinking and expectations in engineering prevent entrance of philosophical aspects into engineering activities. Any engineer should consider that exact solutions are not possible, and hence, the results must be questioned from different facets. For instance, in engineering disciplines the "safety factor" concept kills the possibility of philosophical issues to enter into engineering studies and it is rather an "ignorance factor"; ignorance in the sense that all the blame is thrown over this factor without further ponder and solution in the domain of uncertainty. One can state that the crisper the engineering works, the less philosophical are activities from the start to the end product in engineering.

If one thinks philosophy and engineering as sets of knowledge then without philosophy of engineering today the picture between the two appears as in Fig. 1.2. Is it possible to regard such a situation as dynamic, productive and creative knowledge mechanism? Appreciation of this point is left to the reader.

In current engineering education systems, philosophy and engineering have separate sub-divisions, which cannot have common points, but in reality, one cannot escape from such inference. Logic, epistemology, ontology, art, aesthetics and ethics are all sub-divisions of philosophy. On the other hand, civil, electrical, mechanical, mining, electronic, nuclear, industrial, etc. activities are sub-divisions of engineering. Today, even one can mention about the philosophy of politics, but what about the philosophy of engineering? Are the sub-divisions in engineering arranged according to philosophical principles or according to mechanical principles with crisp and impermeable boundaries? Are there not philosophical and

Fig. 1.2 Philosophy-engineering separations

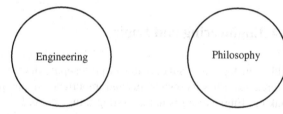

Engineering   Philosophy

logical relations between each engineering domain? The reader can appreciate that at least ethics, aesthetics, art, epistemology and knowledge have relationships with engineering issues. It is neither possible to thing engineering education without philosophy nor to have generative, productive, rational engineering outputs without philosophical ingredients in thoughts. Without the philosophy of engineering and logical rule base in generative engineering career, it is not possible to have engineering outperformances in a society.

In the past, bridges, aqueducts, mosques, cathedrals have been built, say, before at least 300 years without numerical solutions. Is it possible that present day engineering structures can survive many centuries without philosophy of engineering aesthetics, durability, and economic consequences in the long-run? Fantastic examples of such structures can be found today in many places and also in Istanbul, Turkey, as St. Sophia Museum from the fourth century after Christ, and Ottoman Blue Most during the 15th century as well as Roman aqueducts in the center of the city in addition to city walls from Byzantine period. In 1998 there was a severe flood that occurred on the European side of Istanbul in one of the valleys. After the flood occurrence one could see obviously that bridges by Mimar (Architect) Sinan (1490–1588) from sixteenth century were not destroyed, whereas recently made bridges were all demolished or played the role leading to high backflows, and therefore, surrounding areas were subjected to inundation. Ancient engineers and architects did not have scientific tools such as theories, equations, formulations, crisp algorithms, computers and software, but they relied upon their philosophical thoughts, imaginations, linguistic information, and logical rule bases, and consequently, they founded designs on these knowledge sets leading to rational solutions. Present day engineers, without resorting to engineering philosophical concepts, rely on numerical results from well-established equations (are they really so?) or ready software.

Today one may witness many environmental problems, which are almost everywhere due to the absence of the philosophical (epistemological, ethical, aesthetical, logical, etc.) principles and rather blind applications of crisp formulations without interpretations and possible consequent assessments about the aftermath phenomena. One must not forget that any engineering formulation is with a set of assumptions, simplifications and idealizations, all of which may not be suitable for the present problem at hand. These conceptions only trim off the uncertainties in the assessments, and hence, most often drive out philosophical principles. Any engineer with philosophy of engineering and logical principles can suggest alternative formulations to the present potential solution cases. Not static knowledge only, but dynamic philosophy of engineering principles makes an engineer alive, active and successful in his/her career.

1.4 Engineering and Logic

Although logic is a subdivision of philosophy, due to its importance in scientific inferences, it has its own status and importance in any reasoning for final decision making. Engineering is about getting things done, for a definite purpose. Logic is search for rational formal a priori truth, especially, through mathematics, which helps to develop construction and exploitation of abstract or mathematical models. Practical and pure logical principles play emphatic role at the very bottom of mathematical modeling and computer software for rapid, reliable and valid engineering problem solutions. Intellectual content of engineering can be enhanced after philosophical thinking by means of logic rules and principles. Software supporting these intellectual activities is more effective when it is built on solid logical foundations. Scientific and engineering revolutions can be achieved after effective application of logic rules. Increasing rate of logical revolutions is expected in future. In order to take place among such innovative studies logic rules must be deduced for more effective manipulation of knowledge and information.

For many centuries classical logic as suggested by Aristotle has played the main role in many philosophical inferences and scientific methodologies. It is also referred to as crisp (two-valued) logic because there are two alternative decisions (occurrence or non-occurrence; true or false). Numerically true (false) decision is denoted by 1 (0). Crisp logic provided effective expressive power, simplicity, duality, holiness and well developed theories throughout many centuries. Initially, this logic with engineering aspects has been referred to as "logic engineering" and in order to judge the appropriateness of the name, one considers the definition of engineer [3]:

Engineer: one who designs or makes, or puts to practical use.

Reasoning tasks can be achieved by good formal languages, where by "formal languages" has precise syntax and semantics. Reasoning tasks are important when there are inferences and hence a further condition is required on our notion of formal language, namely it should provide a calculus defining some kind of consequence relation.

1.5 Engineering and Science

Scientific research, by its very nature, cannot be reduced to a routine process. Otherwise, the research and technology centers such as universities become mass production plants with mechanistic, dogmatic and no-generic views but with generic certificates. However, these centers would always have much room for improvement in efficiency. Any subject or phenomenon should have the following three conditions for scientific investigation.

- Material, because science is a materialistic system of thought, and therefore, its subject must be material;
- Positions in time, space or in time and space;
- Variability.

The first condition specifies that without consideration of materialistic medium thoughts cannot lead to scientific investigations. Among materials are air, water, soil, metal, electric or magnetic waves, etc. The material must have a geometric shape in space and time. The most important property for scientific issues is that the material should have variations in terms of deformations, movements, dynamism and alike. This is the main reason why in any scientific law there is always change of some quantity with time or space (distance, area or volume). For instance, in the Newton law, the change is the velocity variation by time, which is the acceleration. In Hooks law it is the deformation. In Ohm's law the variation is the change of voltage by time.

Although much knowledge is gained by means book or present day internet facilities nothing can replace experience. The hard way can be transferred to others via the printed pages and linguistic debates. The actual efficiency, productivity, quality and recognition can be achieved through the personal conveyance of knowledge by skilled and experienced personnel who have usually learned many procedures only after years of actual practice. It is necessary to have the hardware facilities such as libraries, laboratories, internet network, periodicals, books and reports, but these are not sufficient to improve the quality in any research center, if experience, skill and quality oriented minds in research and education activities are not active agents.

Basic sciences are prerequisites of any professional training especially in the engineering education. The basic concepts in science must be provided to students in such a way that they give potential births to new ideas and developments. This is only possible through rational and critical discussions of the basic scientific ingredients. Furthermore, in an engineering institution such as the Istanbul Technical University (ITU), additionally, the technicalities must be embedded in succinct and practical manners into the minds of the students. This last stage is specifically important for productive and fruitful expectations of the mind because they are the transition bridges to technological applications, which are among the prime goals of engineering achievements. This stage has significant importance in engineering education rather than sole basic science training. In the past, many scientists have advocated that engineers cannot be scientist but this trend has changed toward the advantage of engineers in recent years. Further changes can be gained by the concept of engineering philosophy and logical inferences about engineering affairs. Especially, in the third millennium engineering education is expected to be more philosophically, logically, ethically, scientifically and artistically oriented.

Scientific research requires originality and creativity and it is very sensitive to psychological state of the scientist and to his/her conceptual sensing of the environment as well as the problems. Science is not a routine process and an uninterested worker is unlikely to produce the new ideas necessary for progress. In any scientific work, there is certainly much room for improvement in knowledge efficiency. Many years of actual practice led to learn different procedures, which make the researcher more skilled and experienced. None of the books can replace experience completely, which is gained through mutual consultations, critical discussions and creative through but the knowledge generated in this way can be transferred to others via publications such as books, journals, reports, lecture notes, etc. However, still preferable way of learning is through linguistic debates and discussions.

Any effort to solve problems, and consequently, to generate useful information and knowledge is referred to be scientific provided that the end knowledge is objective, generative, selective, general and logical. Today scientific activities are still continuing and there is no hope that they will have an ultimate stagnancy or saturation. Hence, anything scientific is in inflation with conquering unknowns and expanding the boundaries of known world by conquering parts of unknowns (metaphysics). Initially, such activities were necessary for the satisfaction of early human requirements for shelter, food and protection purposes. Necessary protective gadgets and opinions were arisen for the common interest of a family, community or nation. These early developments were all in the nature of technological achievements, which led consequently to useful and practical knowledge and information that could be transferred to other communities. In this manner, the support between individuals and communities has started perhaps in an unconscious manner. These early activities and achievements cannot be accounted in the forms of scientific terminology that one understands today. Naturally, these practical achievements and considerations have to be kept in mind, especially in the engineering education, but every effort should be made to select substances that are significant for a wider inquiry pattern. There are numerous scientific achievements so far in different parts of the world and through past centuries due to various cultures and nations.

Today, the level of scientific achievements is interrelated and become the common property of the society. However, presently and more emphatically in the coming centuries, the knowledge will be sold at high costs and even today the retrievals of knowledge and know-how are available at high costs. The ambition of every community, government and nation is to have their own and core individuals that may perform scientific achievements. How could a nation reach to the level of scientific community without training youngsters who will hand over to future generation useful, powerful knowledge and information? This is the main question that is strived for at any cost not only by the individuals, but in a more planned form by the governments and nations. Is it possible to fulfill the scientific knowledge and information for a community by transferring or imitating scientific ideas of other nations? Such an approach has given rise to defective and dangerous prospects among the community and today many governments began to understand that this is not a reliable and proper way for the prosperity of the community. Any community now realizes and completes developments in many aspects of the life, especially in scientific activities, which provide a common basis for other phenomena (social, economic political, military, defense, etc.) in the community itself. It is, therefore, a must that any community who strives for modern and advanced future prospects should have sufficient individuals for the generation of scientifically oriented young minds. These minds are the real core for the development and self sufficiency of any society. This is one of the basic reasons why different countries have different education and training systems in their schools and universities. For instance, the education system is different in USA, England, France, Germany and Japan. Each one tries to develop and optimize with objective criticisms from the instructor to the parents, so as to improve the education system. Especially, traditional, stagnant, static, classical, imitative, and similar non-generative ways are abundant today in most of the education systems.

Scientific work is always rewarding but the most rewarding work is usually to explore a hitherto untouched fields. Unfortunately, today such a task is not easy to achieve. As mentioned above, since scientific works cannot be static, one should always be hopeful for the development of a completely new theory or experimental method or apparatus that makes the scientific activities to enter a new domain. Hence, it is necessary to have more imagination for reaching to such a level. Scientific achievements cannot be attained prior to the successive completion of three steps, which are very common in traditional, say, Turkish philosophy. These are imagination, geometrical conceptualization and inference reflections which are equivalent with the generation of new opinions from available knowledge [24].

Research is an adjective that gives initiative to a scientifically minded individual to search for unknowns not for the immediate requirements but also for long range purposes in any direction. It may also be a mental experience or activity that might lead to any theoretical consequence, which might not have immediate application but waits for some years to emerge as dominant ideas. For instance, a researcher in pure science might not have at all times more problems s/he would like to solve when s/he has time. One may be an excellent experimenter and may have ability to have success in applied scientific activity, but s/he might lack qualities of mind that is a prerequisite for fruitful research. Many branches of engineering fall within the domain of applied sciences and this does not mean that applied scientific activities are inferior to pure sciences, where more mind power and creativity are required. Research oriented scientific mind in any engineering branch may give rise to very advanced technologies and methodologies. Technologies and methodologies require mind activities and functional end products for generative new and up to date information. Orientation plays very significant role and it is the initial ignition power for anybody who wants to achieve scientific works through research. A wise man would know when to abandon a research. It is not possible for somebody to exhaust all aspects of a research topic but there always comes a state where further work with available knowledge is relatively less profitable than the same effort turned towards more fruitful directions.

In engineering, problems are often assigned to research workers by higher authorities, but this does not absolve the engineer from responsibility for examining the statement of the problem with great care. A preliminary condition for any engineer as a researcher to carry out the actual research is to know as much as possible about the background of the problem, how it arose, why is it important, and what will be done with the results?

1.6 Engineering and Modeling

A complex view of reality appears in the form of a scientific model, which is a simplified abstract of the reality. It may represent empirical objects, phenomena, and physical processes in a logical way. Herein, logical way is mostly concerned with rational directly or inversely proportionalities as will be explained in Chap. 4.

The aim of these attempts is to construct a formal system for which reality is the only interpretation. The world is an interpretation (or model) of these sciences, only insofar as these sciences are true.

Engineering is dominated by modeling techniques (physical, mathematical, laboratory) which permit the construction and evaluation of a design prior to physical fabrication of its implementations. In general, models are very useful, but sometimes also dangerous, in particular when the philosophy and logical steps are not well known and they are used unconsciously. Model-based engineering tools for software engineering recognize the importance of architecture and automated analysis with logical foundations and inferences.

The conceptual model implies a model that has conceptual elements. The success of any model depends on its correspondence to a past record, present performance, future prediction with actual state of affairs. These models are usually built by analysts who are not primarily concerned about the truth or falsity of the concepts considered for modeling. All models should have logical propositions as rules or statements with a particular trueness. Mathematical models are representation of conceptual aspects. Any mathematical model is an abstract that uses mathematical language to describe the behavior of a system. Scientific models are representations of physical objects and factual relationships. In any modeling foundation, there are logical statements leading to mathematical expressions by means of various structures such as graphs, clusters, groups, sets, etc. There are even language models, where a structure gives meaning to the sentences of a formal language. If a model for a language satisfies a particular sentence or set of sentences then it is a model of the sentence. Model theory has close ties to algebra, which has first been suggested by Algorithm in the ninth century, who is a Muslim thinker and father of algebra.

Finally another conceptual model is a system model, which describes and represents the structure, behavior, and more views of a system. A system model can represent multiple views of a system by using two different approaches. The first one is the non-architectural approach and the second one is the architectural approach. The non-architectural approach respectively picks a model for each view. The architectural approach, also known as system architecture, instead of picking many heterogeneous and unrelated models, it uses only one integrated architectural model.

Mathematical models are used most often in the natural sciences and engineering disciplines including physics and biology but also in the social sciences such as economics, sociology and politics. Today, physicists, engineers, computer scientists, and economists use mathematical models most extensively. Eykhoff [7] defined a mathematical model as 'a representation of the essential aspects of an existing system (or a system to be constructed) which presents knowledge of that system in usable form'. Mathematical models can take many forms, including but not limited to dynamical systems, probabilistic, statistical, stochastic and chaotic models, empirical and differential equations, or expert systems. Many abstract structures in different models may have interface, and hence, common shares linguistically and logically. For instance, some strength of material, heat transfer, groundwater flow and air pollution problems have the same differential equations, graphical representations but different connotative and abstract meanings convenient for each topic.

A statistical model may depend on a convenient probability distribution function (pdf) for statistically indistinguishable data generation from the actual records of the same phenomena. These models can be categorized into parametric and non-parametric types, where in the former case the pdf's parameters play role, such as the mean and variance in a normal distribution, or the coefficients for the various exponents of the independent variable. However, in case of a nonparametric model the pdf parameters do not enter directly into the model construction but they are only loosely implied by assumptions. In statistics there can be mental (descriptive qualities or physical conceptual in character) event models.

1.7 Engineering and Civilization

Engineering has eye-ball significance since many centuries and even today its preference is more effective in developing countries. First engineers tried to regulate the natural sources and possibilities for the benefit of their community, region and country with care, but after the Francis Bacon (1561–1622) urged as to benefit from the Mother Nature in an unlimited manner, engineers and architects together with many other specialists started to exploit natural resources without much care concerning damage on the nature. Today the size of dangerous damage limits appear clearly in air, atmosphere, water and environmental pollution and contamination problems in addition to misuses of land and soil sources. The measure of societal development of a civilization can be achieved through the existing buildings, roads, bridges, theaters, water supply and distribution networks, arenas, mosques, churches, castles, etc. and their maintenance, which become functional on the basis of views, imaginations, implementations, plans, projects and management of such civilization elements. Experts who try to raise the living standards of their society are among the most important individuals with their generative thinking services, gadgets, instruments and applications in practical life aspects. Such experts are referred to engineers in general. They try to apply simple, economic, easy, fast and practicable ideas after a sequence of trial and error leading to supportive services to the society. In the past, they did not have systematic education but with their intelligence, logic and creative idea generations, they invented methods, procedures, designs and their implementations in an expert manner through gradual experience accumulation.

Engineering affairs did not play basic role only in the development of civilizations but additionally in the economic, social, international, and inter-career affairs. Unfortunately, in our day's internet facilities rendered engineering affairs into more static and traditional appearance as well as into classical forms. It is not frequent that engineering abilities like handy works, eye weariness, mind production and practical abilities are rarely existent. Instead, ready software and speedy communication facilities caused to a significant extent extinction of engineering thinking, creativity, originality, philosophy and production means.

Today even "social engineering" is mentioned, which does not include engineering principles and it implies mechanical aspects of engineering affairs. One can witness that engineering concept has been expanded to cover social and even political areas, but not philosophical and logical rule generations as much. The reason why social and political engineering concepts are coined is because these topics do not include any formulation, equation or crisp algorithm but all linguistic discussions even though they may not be for the benefit of society or politicians.

Engineering as civilization art gives to the society enlightenment and development to reach at top civilization levels. For instance, the term "civil engineering" implies such affairs, but unfortunately, these days, it is conceived as a career that constructs only. Any civilization cannot be without construction of buildings, bridges, dams, airplanes, railroads, express ways, environmental issues, instruments, etc. Such a wide background of activities needs its special artistic, linguistic, philosophic, logical and mathematical means, methodologies and practical solution algorithms dynamically. Engineering does not mean only numerical calculations, but equally important verbal information and knowledge bases. An engineer should mix these two types of information (verbal and numerical) to produce the best services for the society. Unfortunately, today most often artistic side of engineering is overlooked in many activities and only numerical solution sides are given importance for practical solutions. Practicality of solutions also implies technological innovations. Among the main reasons for such a direction is ignorance of philosophy of engineering and extreme attachment to Aristotelian (two-valued crisp) logic, where all the uncertainty ingredients (numerical or verbal) are overlooked through a set of simplifying assumptions. Hence, engineers try to satisfy these assumptions for the obedience to formulations and equations rather than trying to get rid of some assumptions through new and innovative methodologies and practical solutions. As a result of such thought domain, engineers become more formulation, equation, software, algorithm addicted in a static and classical manner. In order to examine many engineers whether the education provided a dynamic or static training, one can ask about what is the significance of Newton's second law? Most often the answer will be as $F = ma$ and in verbal terms, force is equal to mass times acceleration. Such an answer is the proof of static and dogmatic training without philosophy of engineering principles. Philosophy of engineering provides linguistic knowledge and information, which may trigger thought experiments at logical thinking levels. Linguistic knowledge helps to furnish better information generation mechanisms. For example, if one states that the second law of Newton is: *force is directly and linearly proportional with acceleration, provided that the mass is constant*, it will be more understandable even by none specialists. This point shows that formulations are symbolical imprints of linguistic explanations. They are also in symbolic logic forms that must be expressible by propositions, i.e. logical rules. It is also better to generalize the verbal explanation of the Newton law statement as: *two variables are directly and linearly proportional to each other*.

This sentence expresses all the available laws in science. Hence, if two variables are "force" and "acceleration", it implies Newton's second law in physics; electric current and voltage implies Ohm's law in electrical engineering; heat and temperature difference as Fourier law; flux and concentration difference as Fick's law; groundwater velocity and hydraulic gradient as Darcy's law in hydrogeology; speed and distance of a planet from the earth as Hubble's law in astronomy, etc.

It is among the main purposes of this book to explain fundamentals of philosophy as it can be a support to engineering thoughts and creative reasoning together with logical rule derivations all in linguistic terms with examples under the light of scientific principles. The final goal is to train engineering candidates and even graduates in an adaptive manner for productive rationality and self-confidence. The promotion of any society and nation depends on scientific knowledge and technical expertise. In general, science supports the expansion of technology which in turn promotes economic development. Knowledge and wealth have been recognized to be related since ancient times ([2], [14]). Napoleon used to say that: *there cannot be a great nation without great mathematics*.

Today majority of engineers are skeptical about the philosophy of engineering and what it may provide to engineering career. Everything changes including ideas, theories, technologies, etc.; hence, one ponders on the way the change takes place and may ask a set of questions on this matter. The answer to these questions is sought on the linguistic bases, and hence, philosophy starts to play role in rational, critical and constructive thinking. On this grounds intermingling between philosophy, science, technology and engineering may provide additional benefits to engineering career. It is always useful, interesting and important to think about what one is doing, in relation to his/her projects in the social and physical environment. Philosophers or engineers empowered with philosophical principles start thinking about engineering as an abstract thing; they can play a very important role about methods and methodology.

Currently science cannot be performed unless there are engineering gradients in it, and hence, science philosophers should say something about the philosophy of engineering. Philosophers are also concerned with logical structure of explanation issues in addition to the role of uncertainty in the logic of confirmation (Chap. 5). Knowledge also includes belief; therefore, one may ask what is the degree of belief or probability of occurrence of such a belief in actual life [12]. In the past, philosophers attached some probabilistic value to the belief in knowledge but recently degree of belief is another terminology coined in the fuzzy logic context (Chap. 4). Each rule in a set of explanation statements has a degree of belief either theoretically (mentally) attached to it or experimentally if there are actual data [20].

It is not possible to put a clear cut boundary between philosophy and engineering, because if philosophy is driven away from the engineering aspects, then engineer cannot make rational reasoning to reach a conclusive result for the benefit of society. Philosophy contributes to human development on the intellectual and explanation levels, whereas engineering serves humanity for the best comfort in relation to many societal activities. Since, both are concerned with the societal activities, there should be an overlap between the two. Unfortunately, philosophers do not care for engineering affairs in general and engineers think that they are equipped with deterministic knowledge and applicable information, and therefore, they do not need philosophical aspects. Philosophy of engineering is a hidden activity in many engineering issues, rational thinking and logical inferences. It is not possible to distract engineering completely from philosophical reflections, because both are concerned with changes in society and adaptation to such changes for the benefit, prosperity and comfort of society. The humanistic formation of the engineers is bound to improve, reinforcing the interest by the anthropological knowledge, the professional ethics and walking towards advancing social engineering that reflects sensitivity and respect to human beings, society and nature. Such a task can be achieved through an education system where philosophical ponders take place actually in an effective and efficient manner. It is, therefore, necessary to include in engineering curricula philosophy and logic related subjects that help to integrate the social and anthropological contents in systematic forms.

Humanitarian subjects must be in the engineering education system including philosophical and logical courses in relation to engineering aspects, which will be touched in the following chapters of this book. The lecturers should motivate students to criticize any point that is not clear and they must become accustomed to mutual debates and discussions. The students must be educated on different aspects and view directions at looking to piece of knowledge or information. In this manner, one can acquire an ample vision from the critical sense and the freedom of expression. Engineers are also familiar with emotional and human dimensions, which play important roles in the mental formations.

Engineering, whilst it draws knowledge and inspiration from science, mathematics, architecture, art and nature, is neither simply a super nor subset of these disciplines. It has its own distinguishing features. Strangely the discipline with which engineering can best be compared is philosophy or at any rate a modern interpretation of what constitutes philosophy. Adam Morton has stated that [18]:

> Philosophy is one discipline among others, aiming to find truths about the relations between … its objects, in a way that requires evidence from fallible sources, including evidence pre-digested by other sciences. Philosophy is like engineering … concerned above all with topics where theory and evidence are not in perfect agreement, and where practical needs force us to consider theories which we know cannot be exactly right. We accept these imperfect theories because we need some beliefs to guide us in practical matters. So along with the theories we need rules of thumb and various kinds of models.

This puts in a nutshell the very essence that is engineering—to proceed at all, some assumptions or approximations have to be made if 'things' are to be designed and built. There is great art in being able to use gainfully those theories that are known to be imperfect and to judge the extent to which rules of thumb may be safely deployed [10]. On the other hand, Mitcham [17], on reflection, has asserted that:

> ... because of the inherently philosophical character of engineering, philosophy may actually function as a means to greater engineering self-understanding.

and taking this as a lead an increased understanding of the engineer as a global citizen [17]. The same author also points out that engineers are blamed for many of the world's ills (pollution, greenhouse gases, ugly buildings, etc.) and notes that Martin Heidegger

> ... has even gone so far as to argue that all such ethical and aesthetic failures are grounded in a fundamental engineering attitude toward the world that reduces nature to resources in a dominating Gestell or enframing.

The engineer as a global citizen needs to explain him or herself to such a charge! But they need to understand themselves first.

The interest in the interaction between philosophy and engineering has rapidly grown in the last years. Engineering, dealing with the exploitation of scientific knowledge for modeling concrete problems, seems to present several issues worth discussing from a philosophical point of view. Despite many valuable works, a detailed and systematic assessment of the field requires further attention, as it has been recognized from different parts [9]. A commonsense view considers the philosophy of engineering as an area of the philosophy of science, that part concerning in particular the applicative issues of science. This view lies on the idea that engineering is 'just' applied science. Accordingly, the philosophical problems of the philosophy of engineering would be 'just' the problems of analyzing the passage from theory to application, as if a clear-cut distinction between science and engineering should exist. In conclusion, philosophy of engineering, while sharing a great deal with the philosophy of science, also presents some peculiarities, which are worth stressing in the effort of a further assessment of the field. These peculiarities concern both the problems, which are related to the modeling activity typical of engineering and, thus, are more concrete than in the philosophy of science, and the method enriched by experimental verification.

There are good arguments for considering engineering from a philosophical point of view with specific reference for example to empiricism, rationalism, existentialism, logical positivist, post-modernism, and the philosophy of science. The way engineers interact together can be interpreted from a philosophical standpoint and a similar treatment but with an external focus (e.g. dealing with engineer non-engineer relationships) can be applied to the external perception of what constitutes engineering. When taking what might be termed a holistic and philosophical perspective some conclusions can be reached that suggest that the engineering profession needs to partially realign itself away from a purely scientific base in addressing the major challenges facing humanity today. The underlying reason is that engineering is not just science—it may use science and clearly science is of huge importance to engineering—but it is much more and needs typically to take into account a wide range of factors and aspects. So for that reason this author, at least, dislikes the use of the term "Engineering science" as it carries the suggestion that Engineering by itself does not embrace science! Finally, as a means of communication the Engineering profession can utilize the tools of philosophy to help enhance the understanding of all citizens regarding how engineers come to their conclusions and solutions [10].

References

1. Aleksander I, (2006) What is engineering? The Royal Academy of Engineering, Philosophy of Engineering, Monday, 27 March 2006; pp 2–6. Accessed on Dec 15th. Available at http://www.raeng.org.uk/policy/philosophy/pdf/Transcript_of_Presentations_on_27_ March.pdf. (Accessed on Dec 30, 2007)
2. Condorcet MJA (1794). Sketch for a historical picture of the progress of the human mind: tenth epoch. Daedalus 133(3):65–82
3. Davidson G, Seaton M, Simpson J (1994) The Wordsworth concise English dictionary. Wordsworth reference. Wordsworth Editions Ltd., Hertfordshire
4. De Poel IV, Goldberg, DE (2010) Philosophy and engineering. An Emerging Agenda. Springer, Dordrecht, 361 pp ISBN 9789048128044
5. Davis M (2005) Thinking like an engineer. Studies in the ethics of a profession. Oxford University Press, New York
6. Engineers' Council for Professional Development (1947) and Canons of ethics for engineers. McGraw-Hill Encyclopedia of Science and Technology (1982. McGraw-Hill)
7. Eykhoff P (1974) System identification: parameter and state estimation. Wiley and Sons, London
8. Goldmann S (1990) Philosophy, engineering and western culture. In: Durbin Paul T (ed) Broad and narrow interpretation of philosophy of technology. Kluwer Academic Publishers, Dordrecht, pp 125–152
9. Goldmann S (2004) Why we need a philosophy of engineering: a work in progress. Interdisc Sci Rev 29(2):163–176
10. Grimson W (2007) Engineering—an inherently philosophical enterprise. In: Cristensen SH, Meganck M, Delahousse B. (eds) Philosophy in engineering. Academica, Aarhus, pp 89–102
11. Hacking I (1967) Slightly more realistic personal probability. Philos Sci 34:311–325
12. Keynes JM (1921) A treatise on probability. MacMillan, London
13. Luegenbiehl HC (2010) Ethical principles for engineering in a global environment. In Philosophy and Engineering. In: De Poel, IV, Goldberg DE (eds) An emerging agenda. Springer, Dordrecht, pp 147–160. ISBN 9789048128044
14. Marshall A (1890) Principles of economics. Amazon Kindle Edition
15. Meijers A (2009) Philosophy of technology and engineering sciences. handbook of the philosophy of science. 9. Elsevier. ISBN 9780444516671
16. Mitcham C (1994)Thinking through technology: the path between engineering and philosophy. University of Chicago Press, Chicago, 397 pp. ISBN: 0226531988
17. Mitcham C (1998) The importance of philosophy to engineering. Technos vol XVII/3/http:// campus-oei.org/salactsi/teorema02.htm
18. Mou B (2001) Two roads to wisdom?: Chinese and analytic philosophical traditions. Open court, Chapter 3: philosophy as engineering, Chicago
19. Mitcham C, Mackey R (2010) Comparing approaches to the philosophy of engineering: including the linguistic philosophical approach. Van de Poel, I, Goldberg DE (ed) Philosophy and engineering. An emerging agenda, Springer, 49–59

20. Ross TJ (1995) Fuzzy logic with engineering applications. Louis St , San Francisco (eds) McGraw-Hill, Inc, New York
21. Scharff RC, Dusek V (2003) Philosophy of technology: the technological condition. An anthology. Blackwell Publishing, Oxford. ISBN 978-0-631-22219-4
22. Seely BE (1999) The other re-engineering of engineering education, 1900–1965. J Eng Educ 88(3):285–294
23. Şen Z (2011a) Mühendislikte Felsefe, Mantık, Bilim ve Etik. (Philosophy, logic, science and ethics in engineering). Istanbul Technical University, 216 pp (in Turkish)
24. Şen Z (2012) Engineering science and philosophy. International Res J Eng Sci, Technol Innov 1(1):14–25
25. Vermaas PE (2010) Focusing philosophy of engineering: analysis of technical functions and beyond. In: Van de Poel I, Goldberg DE (eds) Philosophy and engineering. An emerging agenda. Springer, 61–73

Chapter 2
Intelligent Reasoning Elements

2.1 General

Human can understand objects through sensory organs and then translate these feelings into some imaginative notions, geometries or figures, which can be subsequently explained to other individuals so that they can also experience similar feelings, and if necessary, make their comments to reach a common objective inference. The root trigger of these stages is the philosophy, which may be expressed in plain terms as the "love for knowledge". Similar to someone who falls in love with some object, s/he then imagines, designs (describes) the feelings, and finally, states them into formal words, oral sentences and into written text.

Human beings are creatures, who can think and take decision for daily life activities towards better prosperity. They are even referred to as "clever animals", which can judge the circumstances and try to reach the goal whatever it may be. Five sense organs provide information from the surrounding environment, and accordingly, the decisions are taken after the logical and rational judgments. However, since the origin of life for many centuries, the judgments are internally processed by human mind and results are put out. Engineers are not distinct from such human mind activities, but the current engineering institutions mostly do not care for the philosophical and logical principles, and consequently, majority of engineering graduates are addicted to algorithms, mathematical equations, formulations without the basic foundations, but they try to apply what they have learned as a prentice without or enough reasoning. If they do so, then many present day formulations, algorithms will evolve by time towards better advancement levels. The fundamentals of any knowledge and information are language expressions of all the imaginative, descriptive and generative ideas, which evolve through the sequence of root words (terminology), sentences (propositions) and informative texts (reports, books, papers).

Z. Şen, *Philosophical, Logical and Scientific Perspectives in Engineering*,
Intelligent Systems Reference Library 143, DOI: 10.1007/978-3-319-01742-6_2,
© Springer International Publishing Switzerland 2014

2.2 Word Roots (Etymology) and Meanings (Epistemology)

The scientific explanations are normalized views in natural language, which codes any type of knowledge and information in words and dynamic way in sentences. All inferences at their earlier stages are linguistic expressions, which drive mind towards a certain direction for meaningful inferences. All scientific laws are expressed linguistically in terms such as the Newton's second law states that "there is a directly proportional and linear relationship between the force and acceleration", which is then onwards converted to mathematical symbols as $F = ma$, after logical deduction of such a relationship. Scientific explanations are primarily in the forms of deductive arguments, from which scientists deduce predictions and then try to verify whether those predictions are correct. If some of them are not, the hypothesis is disconfirmed; if all of them are, the hypothesis is confirmed and may eventually be inferred. Unfortunately, while this model does make room for vertical inferences, it remains, like the enumerative model, far too permissive, counting data as confirming a hypothesis which are in fact totally irrelevant to it. For example, since a hypothesis (H) entails the disjunction of itself and any prediction whatever (H or P), and the truth of the prediction establishes the truth of the disjunction (since P also entails H or P), any successful prediction will count as confirming any hypothesis, even if P is the prediction that the sun will rise tomorrow and H the hypothesis that all ravens are black [15].

It is possible to determine rationally the value of any notion in words, which leads to grasp of the idea. This statement indicates that the best medium of understanding is the native language. Even though there may be a series of graphical, equation or symbolic presentations of the idea, the words are the essential expressions and symbols for basic understanding. Language helps to express one's thoughts, which are translated into words and then by audio-visual sense organs one can understand the refined idea. In practice, the more plane is the language the more one can express the idea clearly towards better understanding. Systematic thinking is a disciplined way of understanding; otherwise confusions enter the thinking process without any objective debate. Mother tongue has many traditional meaning reflections in words, spelling and grammar of that language, so any deflection in the tradition may lead to communal misunderstanding. Philosophy needs language more than anything for meaningful understanding.

The historical origin of words lies in etymology, which searches for the evolution of each word meaning content and form. Most of the words are related to local objects or events for their origin and also they may be adopted from other languages. Some of them are driven by adding prefixes and suffices and also through transliteration. Many words become obscured by time due to either sound or semantic changes.

The branch of science that searches for the meaning as knowledge and understanding is epistemology, which is one of the most important branches of philosophy. It searches for the answer of the questions as what is knowledge. How is knowledge acquired and to what extent is it possible for a given subject or entity to be known? Proper answer to these questions can be obtained after identification of nature of knowledge and its relationship with truth, belief and justification. A detailed account about the truth, belief and vagueness is presented by Machina [13]. The theory developed by Edgington [6] holds that vagueness is not a special semantic phenomenon, but a consequence of the nature of linguistic knowledge and general principles of language use. The precise development of this claim, however, will not be in terms of "fuzzy" regions but in terms of probabilistic linguistic representations. There is independent reason to believe that human knowledge is represented probabilistically. The scope of human knowledge with its scope and limits are explained by Russell [18]. Lassiter [11] thinks that a perspective leads naturally to a model of interpretation as an interpreter mapping words and other utterance-types to a probability distribution over precise resolutions.

In this book, linguistic impressions and expressions are advocated for idea generation, and therefore, it is necessary to understand words, sentences and propositions for better philosophical and logical activities. Humans are capable to generate knowledge by conversation and mutual agreement or disagreement. For success, it is necessary to know, what are the epistemological contents of each word? In the meantime, jargons, confusions and rather than common and joint understandings, individual grasps and interpretations may lead to debatable conclusions, which may open further discussion opportunities. The root meaning of each word has importance in content interpretations. Epistemology as a branch of philosophy implies knowledge and meaningful understanding. This needs analysis of nature of knowledge and its connections to truth, belief and justification. In short epistemology is concerned with the theory of knowledge down to the root. The main purpose is to search for holistic meaning of each word as for its linguistic information content. Clear understanding of each word leads to the theory of knowledge or information, which provides one to construct logical propositions. A set of logical statements explains simply the whole of any phenomenon. Words gain meanings by examination of objects in a rational manner. Rational naming of each specification by a single word can be obtained on hearing from others during conservations, and hence, each specification, adjective and other properties of all the objects gain linguistically meaningful knowledge. Hence, epistemology helps to locate information content of each word in one's mind. In thought system, linguistic explanation of any equation, formulation, concept, plan and project, knowledge play initial role.

In engineering investigations this continues until the mathematical formulations of linguistic expressions leading to numerical solutions. Linguistic information is drawn from the ocean of philosophical thinking, and especially from the convenient section of engineering philosophy. The meaning of each word must provide automatic animation in the mind for sound and meaningful perceptions. Naturally, such an animation may take some time for full automation in order to give information content of the word. After frequent repetitions of the word perception, mind recognizes it with its epistemological content for use in daily life and during any related research.

In any language, words gain content and meaning in minds after audio-visual practices, and hence, they become common perception tools in the society.

Anybody can grasp the meaning of each word because s/he talks everyday his/her language, and hence, automatic and unconscious meanings appear on his/her mind instantaneously. Such automation helps one to express himself/herself with least misunderstanding in many conversations, debates and discussions. In cases of misunderstanding further explanations through other set of words help to minimize the bridge of differences between individuals' understandings. Each word describes various internal and external properties of abstract and concrete objects. Words are symbols of such properties not in a crisp manner but with ambiguousness, incompleteness, uncertainty and fuzzy content in most of the cases. Objects around us do not imply any meaning without human mind and perception. Words are the conveyors of different aspects of any object to human mind for grasping, understanding leading to subsequent mind treatments. Like naming of a newly born baby, object properties are also named for their simple recalls in the mind. Words help to preserve information and knowledge in our minds. One takes the initial meanings in mother tongue and s/he may later, if necessary, translate the same information or knowledge to other languages. One can recognize the object concerned through the word information content and takes position against any situation according to its meanings. For instance, "cliff" as a word implies very dangerous situation that one should absolutely avoid for his/her life protection and sustainability.

2.3 Sentences

In daily conversations any expression in native language is a sentence, which has grammatical pieces of few words each bearing syntactic relation with other neighboring words. Generally, there are three types of sentences as simple, compound and complex. A simple sentence is an independent clause and contains a subject and a verb. Subject and verb are sufficient to express a complete thought about an object or event. The following sentences are simple in form but imply important thoughts.

- Plumb is heavy;
- Fishes swim in water;
- The house was tall.

Notice that each one of these sentence includes some components of scientific criteria as materials (plumb, fishes and house) and specifications (heavy, water and tall). Each one creates a thought in our mind.

A compound sentence contains two independent clauses joined by "and", "or", "nor", "for", "yet", "but" or "so". Conscious use of each one of these 7 words can change the relationship between the clauses. Few examples for compound sentences are,

- Groundwater is replenished by rainfall or after snow melt;
- Gold is expensive but it is also very precious;
- Atmospheric movements are necessary for wind power.

the computer memory and their call through a program for some pictures on the screen, likewise concepts when called from the memory storage provide such pictures on the mind screen. Knowledge that is in the mind storage can be used in daily life and in any scientific and engineering activity process. Everybody has concept storage in his/her mind that may or may not be common with other individuals. In a way, design of concepts by thinking, is in the form of quiet talking. Concepts are embodied in the mind as imaginations. They cover different generalizations according to the persons and objects. For instance, "tree" concept implies that it includes general tree information without making any distinction between tree types. With such a concept in his/her mind anybody can write a general composition about the tree. The imagination of "tree" concept in our minds is not a single tree. Scientific knowledge has generalization property and this is also an example for the concepts. Imaginations in creating concepts in human mind are general. Otherwise, imagination of someone about his/her lover is not a concept, because it is an individual case, subjective and not general.

Concept perception must emerge from the mind and then given to communication system. The single most important communication system is the language, which is a must for the appearance and crystallization of concepts in human mind in the forms of different shapes and then become ready as information content for use and further knowledge generations. Even through the concepts are in the form of abstractions in human mind, they become more concrete through the language as sayings, symbols as writings and as pronunciations. These symbols are referred to as terms, which are the smallest pieces in any communication means. Hence, one can understand the importance of concepts and terms in our thought system for perception and communication. Prior to any study, and especially modeling in engineering, various concepts related to the investigation topic must be judged rationally on philosophical bases (Chap. 3). In fact, any education system should adopt this trend for fruitful productions (Chap. 6). Each topic has its special basic concepts and terms. It is also possible to generate new concepts after thought experiments. Collection of all the related and meaningful terms about a topic is referred to as terminology of that topic. Hence, terminology is a dictionary of the words, concepts and especially terms related to the topic concerned. Terminology of a topic and its design in mind through thoughts and concepts must be understood correctly. Each word with its meaning can be considered as a term such as tree, water, fire, Turkey, come, go, read, etc. However, words like "and", "or", "not", "however", "nevertheless", etc. are not terms.

After understanding of concepts and terms, one of the human thought ingredients is definition. Each definition helps to understand temporal and/or spatial features of the object or phenomenon. Definitions should be simple with practical information contents as meaningful collection of words, i.e., in the form of a complete sentence. Combination of concepts in sentences provides knowledge about the quantity and quality of objects. There are many definitions in various disciplines. Generally, science and engineering definitions show quality or quantity variations per unit time or space. According to this, for instance, in physics speed is the length covered in a unit time. Again in physics, work is defined as force multiplied by distance. Power is another definition as the work per unit time. Here, length and time are terms and they have information contents in mind. For instance, volume of water during a unit time is another definition named as discharge.

2.5 Language

Knowledge gain starts with vague perceptions and the more one tries to describe the object at the focus by statements, the more s/he knows and explores its appearance features. If s/he is unable to put the thoughts into words, then there are restrictions for dynamic understanding. Hence, any language is not merely the sole medium of thought, but in the meantime, it is the very stuff and process of dynamic thoughts and their evolution towards completion. The words are necessary for describing the geometry, color, taste, hardness and possible sound, which are all static descriptive features of the subject. The specifications do not change by time and they have different wordings in different languages and through the linguistic translations; everybody comes to the same understanding about the subject. Any notion should be expressed in words for appreciation of the value of such impressions. Without any notion, appreciation and impression the generative idea will remain obscure, but otherwise, the final statement has rational content. Clear expressions in words help anyone to know the basic idea, and hence, language becomes the essence of understanding. Sketches, graphs, diagrams, pictures and drawings help to support each word even though one does not have in front of him/her the physical object, because the meaning in the word will revive in the mind the same object in an imaginary medium. Language helps one to understand objects or events and the following points are among the most significant roles of any language.

- Thoughts are loaded onto the linguistic expressions, which aid to translate thoughts into words for refined understanding;
- Plainly simple words help to understand the idea, and hence, the root of such words is related to some physical, chemical, mechanical, psychological, etc. characteristics of the object. The shorter and more descriptive the words, the better is the understanding;
- Thoughts on the meaning of each word improve understanding, and therefore, native tongue is the key for direct grasp, perception and understanding;
- Plain language reveals the idea more easily, clearly and the more clearly revealed an idea, the better becomes its understanding, otherwise, it cannot be understood clearly.

Languages have evolved into disciplines with rules of spelling and grammar for better understanding of meanings and the relationships between various objects or events. This premise indicates that clear and obvious thinking and thought generations should abide with a set of rules and regulations, otherwise it is not possible to understand the ideas. Hence, the crucial importance of mastery of language use should be relevant both to individuals and to community. The only way such mastery can be maintained is by regular exercise; that is by frequent discussion, reading, writing or speeches. However, in the scientific arena, sentences that have rational relationship content are sought for further idea generation or better understanding of the ideas.

2.6 Knowledge and Types

In general propositional sentences convey knowledge, which may have explanatory, question (concerning how) or acquaintance types. Today "know-how" is an expression used for a bundle of useful knowledge to answer "how?" type of questions for practical solution of issue concerned. There is also theoretical knowledge that needs to be searched further for confirmation through experimentation or verification by factual data. Epistemological content of any proposition under the light of logic rules and principles brings out the meaningful and useful knowledge for practical uses or for further idea generations (Chap. 4). In the meantime intellectual capacities and virtues of individuals in general and specialists as engineers in particular can also be evaluated by epistemological considerations. It is not only through formal education and training that one gains knowledge and renders them into useful information forms, but also life experiences of each individual open ways of another knowledge source, empirical knowledge. Here, perceptual observations and the senses play important role. Such knowledge can be expressed after the full meaning of the content in terms of first logic and then mathematics in physics and engineering. Another source of knowledge is acquired by a priori processes (innate) that are not derived through experiences. Innate human knowledge is rather subjective varying from individual to individual depending on his/her emotional feelings, ambition to acquire knowledge, interrogation abilities and opportunities. It is not possible to draw a crisp border between innate and empirical knowledge types. Vague border existence between these two types provides availability for interaction towards new knowledge generations.

If knowledge is regarded as the basic stone of a society or an individual in knowledge production, this is a general definition of the knowledge from benefit point of view. Improvement and ripeness processes of knowledge take place through systematic accumulation of human philosophy and science, which are missing essentials in engineering education (Chaps. 5, 6). In general, philosophical thoughts render abstract knowledge into more concrete forms, and finally, even though they may not be complete but fuzzy due to uncertainty (verbal or numerical). These are referred to as philosophical knowledge. Since almost three centuries, entrance of these knowledge types into science arena, which requires objectivity, generality, logicality, selectivity, falsifiability and unbiasedness, they become to be regarded as scientific knowledge. Scientific knowledge generations require certain methodologies and systems. These are of the type that are not subjective and find their location in any society irrespective of the cultures and engineering knowledge is of such type.

Potential knowledge sources are internal human senses, the society and relationships in it as well as natural and daily life phenomena. In order to sustain daily life, one needs to have practical knowledge; for investigation of natural events through methodologies and their end products as positive knowledge; and like justice, economy, psychology, politics, etc. as social knowledge.

Knowledge that is used in a useful manner for the mutual trust, comfort and security in a society must be purified towards more beneficial direction. Beneficial facets of any knowledge can be measures through niceness and goodness, which are rather relative concepts. Thus knowledge is relative not absolute. Since absolute niceness and goodness are not known, human beings try to improve their knowledge for betterment as time passes. During this process one may come across with knowledge that is better than previous ones. Niceness and goodness criteria help one to select the best knowledge level at any time. Especially, nice and good behaviors open the ways towards honorable feelings in the communication world. First Greek philosophers like Socrates regarded knowledge and honor higher than ignorance and crookedness, and hence, they tried to correct wrong knowledge in the society. Plato, another old Greek philosopher, though that pure and absolute beings are in the form of "ideas". However, Aristotle accepted knowledge as controlling mechanisms of human behavior. He advocated that if a man knows niceness and goodness then s/he can control various behaviors towards an honorable direction.

Up to now beneficial, good and nice knowledge are preferred for healthy development of a society. However, this does not mean that anybody with goodness and niceness properties cannot be honorable. One can generate also ugly and bad knowledge, but their unused by anyone makes him/her more honorable. Distinction of nicety and goodness can be achieved by thought and its engine is mind, and thus human can direct internal wishes towards better direction as an important difference from animals. Consequently, Aristotle considered humans as "thoughtful and thinking animals".

Knowledge is objective and generative if they connect human to his/her environment and explain events in a clear and selective manner. Their perceptions are realized by human sense organs. This knowledge is then transferred into the brain in a vague, incomplete and suspicious manner, and hence, they ignite the thoughts of individual. If this knowledge is stored in the memory as individual knowledge, they are called as "perception knowledge". These can be withdrawn from the memory storage and transferred directly to other individuals.

Use of the perception knowledge and memory stored knowledge withdrawals automatically and frequently renders any individual to more experienced modes. This also provides transformation of perception knowledge into mentally experimental knowledge, provided that the thinker cares for philosophical and logical ingredients linguistically. After such a mechanism one can transfer the knowledge to other individuals practically and automatically in detailed, open and clear manners. For example, if someone sees somebody, who rub two very dry woods together and obtains fire, then s/he can transfer this perceptive knowledge to others. Since s/he did not care for the experimentation of this event, detailed knowledge transfer is not possible. On the other hand, one who made such fire many times gains detailed information as to the shape, dryness and type of the wood in addition to the force of rubbing. These are now experimental knowledge, which are very beneficial for the society at large. This indicates that perception knowledge usage frequently gives rise to experimental knowledge. Anybody with experimental knowledge can answer questions of why the event takes place, but not yet how it works? In order to answer this question after perception and experimentation, one has to ponder philosophically by asking himself/herself continuously at every stage of the event evolvement, how? and philosophical, rational and logical answers can be given. Hence, another set of knowledge is obtained, which is referred to as "art knowledge". Still we do not have scientific or engineering knowledge because the event has not been formulized for numerical calculations. It is not necessary that one should go through a systematic education to arrive at art knowledge, but systematic and critical reasoning for identifying relationships between various stages of the event is enough through perception and experimental knowledge.

According to Aristotle, in addition to three knowledge types mentioned above, there is a fourth one, which involves reasons of event occurrence and their preliminary explanations, which is philosophical type of knowledge. Anybody who reaches this level of knowledge is a philosopher or gains philosophical experience to attack the problems for solution. He uses the three knowledge types in search for reasons and explanation explorations and tries to reach a more detailed knowledge level linguistically. Philosophical knowledge is among the most teaching means to human and it helps to generate further knowledge. Since history immemorial, philosophers guided many thinkers towards right directions in their specialization, and hence, after some time science, art, technology and engineering became separate from the philosophy on their own knowledge generation channels. However, especially engineering become very independent from philosophical knowledge, and hence, instead started to encourage formulation, equation, algorithm, software, etc. Mechanical knowledge without any inert explanation kills philosophical reasoning, even perception, experimental or artistic reasoning. This is the main reason that any engineering education and engineer will need philosophy of engineering principles and fundamentals to become more fruitful, productive and generative in their career (Chap. 3).

Engineers become specialists that give attachment to practical application of any knowledge type without development of his/her knowledge memory in a dynamic way. They do not care for philosophy but only for end products towards practical applications. Philosophical knowledge generations do not care for whether they have practical applications or not, but knowledge is obtained most of the time for the sake of better knowledge improvement.

Another version of scientific knowledge is the set that helps to satisfy daily requirements of individuals as human health, justice, construction and many instrumentation affairs that require engineering works.

2.7 Human Mind

Our mind is the generator of uncertain impressions and conceptions. It divides the seeable environmental reality into fragments and categories, which are fundamental ingredients in classification, analysis and deduction of conclusions after labeling each fragment with a "word" such as a name, noun or adjective. The initial labeling by words is without any motion and without interrelation between various categories. These words have very little to do with the wholeness of reality—a wholeness to which all of us belong inseparably. Hence, common words help to imagine the same or very similar objects in our minds. Furthermore, the real world is pieced together from fragments, which are made out of sensations, thoughts and perceptions. They serve collectively to provide partial, and therefore, distorted conceptual models of reality, which represent perceived human-mind-produced world. It is not a world whose natural evolution has brought us to existence and with which we are linked through an umbilical cord of vital and impossible-to-separate connections [4].

All the conceptual models deal with parts of something that is perceived by human mind as cover-there, as surroundings considered to be used for what our ego-centered minds label as meaningful. Of course, among these meaningful fragments there exist clear and hidden interrelationships, which are there for the exploration of human intellectual mind. Unsupervised or supervised (trained) minds on any topic such as scientists, engineers, economists, politicians and philosophers are involved in adapting many distorted conceptual models for predicting and exercising power over unfolding dynamics of reality, which are beyond our ability to predict and control absolutely. However, scientists and eager minds try 'to do their best' to mutilate reality so that it could be pushed into meaningful beds of senseless model reductions. In doing so, the human mind is dependent on the fragmental information that could be gathered about the reality.

The mind confronts with dilemma or duality, and hence, it either selects something while rejecting its opposite. This trains the mind in black and white crisp thinking as a first approximation to model the reality (Chap. 4). Such a distorted model of reality based on duality is referred to crisp or binary logic, the foundation of which was established by Aristotle, who lived around the third century before the Christ. Although prior to Aristotle human mind was based only on natural and innate logical principles, but it became restrictive with the duality principle preference. The dualistic nature of rational reasoning component of mind is so strong that mind alone is unable to transcend it; the best it can do is to reconcile the opposites. Hence, the crisp logic has no vagueness, ambiguity, possibility or probability because everything is either white or black. Keefe and Smith [10] mention a clear and very useful overview of theories of vagueness, and they collect many classic papers on the subject. Classical black-and-white approach in thinking can easily entrap human mind in routines, stereotypes, prejudices and habits that become a source of fuzziness, which eventually makes one incapable for authentic experience. This is because all our 'understanding' is constantly filtered through already establish mental patterns. Fanaticism is an extreme manifestation of this kind of dense fuzziness, when human ability to move beyond an established dogma is entirely blocked.

On the other hand, even today as human beings we all have vague, ambiguous, uncertain, possible and probable concepts and approaches towards our daily affairs. This natural logic is wider and more general than the crisp logic, and therefore, it is labeled as fuzzy thinking when using fuzzy or probabilistic reasoning, where it is possible to accept both the opposites up to some degree of belongingness. By following the fuzzy logic-based approach in thinking one can agree with everything the others say and this can easily push us towards compliance and indecisiveness. When everybody is right, the uncritical acceptance of the fuzziness accompanying other people's thoughts makes it hard for one to generate his own creative ideas. The polarity of opposites, contradictions and clashes of opinions provide human mind with dynamics necessary for transcending the opposites. These dynamics manifest in mind as an urge for searching beyond the plane where the opposites clash and without such an urge, the mind can be entrapped into static, stuck in repetition or memorized by illusory thoughts and dreams [4].

For creative research with fruitful and innovative conclusions one is advised to be able to go beyond the established classical logical rules and restrictions no matter how soft (fuzzy, probabilistic) or hard (crisp, binary, deterministic) the concerned phenomena are. As far as the process of revealing "nature's best secrets" never stops, what "we think we knew" yesterday inevitably changes today, and new vistas "whose splendor we had not even come close to imagining" constantly open to those who are thirsty for knowing. The fuzziness of knowing never ceases to exist. This is a paramount characteristic of the human knowing, which challenges humanity and constantly propels its search for truth and understanding the secrets of reality.

2.8 Thinking Stages

Any scientific thinking has three major steps, namely, imagination, visualization and idea generation. Figure 2.1 indicates the steps necessary in a complete thinking process [20]. Each one of the steps cannot be explained in a crisp manner and each individual depending on his/her capabilities may benefit from this sequence.

The imagination step includes the setting up of suitable hypothesis for the problem at hand and the purpose of the visualization step is to defend the representative hypothesis. Scientists usually use a variety of representations, including different kinds of figures (geometry) to represent and defend the hypotheses. Scientific hypothesis justification is possible only through the understanding of visual representation, and if necessary, modification of the hypothesis should be in progress. On the basis of hypotheses, the scientists behave as a philosopher by generating relevant ideas and their subsequent dissemination, which should include new and even controversial ideas, so that other scientists can surpass and further elaborate on the basic hypotheses. Whatever the means of thinking are, the

Fig. 2.1 Thinking gradients

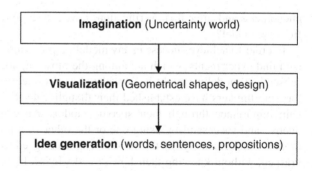

scientific arguments are expressed by linguistic expressions prior to any symbolic and mathematical abstractions. In particular, in engineering, the visualization stage is represented by algorithms, graphs, diagrams, charts and figures, which include a tremendous amount of condensed linguistic information.

Scientific visualizations have been conducted with geometry since the very early beginning of scientific thoughts. This is the reason why the geometry was developed and recognized by early philosophers and scientists over any other scientific tool (such as algebra, trigonometry, or mathematical symbolism). Al-Khawarizmi (died 840 A.D.) who is known in the West by his Latinized name "algorithm" solved second order equations by considering geometric shapes [22]. For example, he visualized x^2 as a square with side equal to x, and any terms such as ax is considered as a rectangle with base length x and height equal to a. This geometrical thinking and visualization made him the father of "algebra". All his discussions were explained linguistically (Chap. 4).

All conceptual models deal with parts of something that is perceived by human mind. Among the meaningful fragments of the phenomenon, there may be clear or hidden interrelationships, which are for further exploration of human mind. Fragments of thinking, sensations, thoughts and perceptions serve collectively to provide partial and distorted conceptual models of reality in representing a perceived human-mind-produced world.

Although human wonder and minds are the sources of uncertainty in forms of vagueness, dubiousness, incompleteness they also serve to overcome problems through human experience, expert views (Chap. 5). The uncertainty concepts in understanding complex problems are dependent on observations, experiences and conscious expert views. When problems are solved, there is always remaining uncertainty that paves the way for future developments. Thus, scientific solutions cannot be taken as absolute truths in positivistic manner.

2.9 Thought Models

Let us think to answer the question, "how an engineer attaches significance to innovations during and after the systematic education? Among the main subjects of engineering is the establishment of suitable models for investigations of real phenomena and after their verification, how future descriptions can be obtained from the model?

In order to achieve success in any modeling procedure, it is necessary to visualize mind experiments, which are among the most important gadgets. Thought and subsequent modeling might be considered as two faces of a coin [19]. In the history, big thinkers have established their thought rules by use of models. Thoughts gain importance through their specific models. An engineer may become productive and successful by using one or the other or a mixture of various thought models. Unfortunately, in the present engineering education system models are grasped, without knowing their how's? and why's?, by memorization and without critical discussions. Such a tendency become more frequent due to the increase of software as a result many engineers adapt their works in a robotic (blind) manner without knowing the flow charts or functional philosophical or logical fundamentals of the software. This may be one of the main reasons why today majority of engineers may not become productive during their systematic education and/or after graduation (Chap. 6).

Any individual during his/her investigation of surrounding nature, relationships with other individuals in the society or phenomena that are in relation with himself/herself starts by benefiting either from simple and small knowledge accumulations or classifications leading towards a higher level of information and knowledge content. On the contrary, it is possible to make the investigation from a whole by partitioning to small pieces towards the final destination as to know "why's" and "how's". In general, the first model may be called as top-down and the other down-top procedure. In the formal literature, the first thought model is referred to as "deduction", whereas the second alternative is the "induction". The same event or phenomenon can be investigated by either one of these models or better by both of them in a hybrid manner, which provide inputs, outputs in addition to the generation mechanism of the whole process. Figure 2.2 provides descriptive representation of these two models. In both systems, the thought box in the middle includes mind functions, where philosophy and logic play the sole role leading to various logical, analytical, statistical, empirical, algorithmic, mathematical, etc. models.

Induction is the education model that is applied in different countries as though model that many philosophers, researchers and educationist commonly adapt. In this system, the students take knowledge in pieces and with rational collection, and

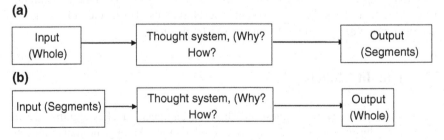

(a)

| Input (Whole) | → | Thought system, (Why? How? | → | Output (Segments) |

(b)

| Input (Segments) | → | Thought system, (Why? How? | → | Output (Whole) |

Fig. 2.2 Thought models **a** deduction; **b** induction

hence, they are urged to reach to higher productive knowledge levels. In addition to the basic information in his/her area of domain, the student must also take additional information in related topics so that they can manage to collect all pieces and combine them in a systematic manner with proper understandings of "why's" and "how's" in order to reach to whole forms of information. Among the related topics philosophical and logical information lead to the list of knowledge even though the students may not know them as an expert, but some sufficient knowledge about these topics, helps them to achieve their final goal with achievement. These are the main topics, which provide rational answers to "why's" and "how's" of a problem and its solution.

Unfortunately, current education systems, like many disciplines including engineering, the students are trained in such a way that there are not leaky compartments among various disciplines as shown in Fig. 2.3.

This leads to a wrong impression as if there are not common interests or points among different disciplines. The necessary interactions among the various disciplines can be provided by the philosophical aspects and its subsequent logical rule bases. If the philosophical and logical aspects are not given in a sufficient dose to engineers during their graduate studies, then inductive type of educational system fails to empower the youngsters for creative and productive innovative ends and such a system becomes rather unproductive. In such a case, many engineering students' creative skills are weakened without useful outputs. Of course, memorization and knowledge transfer without reasoning enter the circuit of training. This leads further to standard prototype engineers without any preference of one over the other. Like leakages among various disciplines, even within the same discipline different courses are confined in themselves individually without any communicative channels with other subjects.

In the deduction thinking system after the explanation of any event to a set of individuals, everybody might understand and partition the wholeness into a set of pieces, and even by identifying the relationships between these pieces; it is also possible to reach inductive conclusions. The major problem in any engineering education system is to construct models for problem solutions unfortunately without consideration of thinking models. Due to such deficiency, engineering solutions remain frequently in the confinement of memorization, knowledge transfer and application without reasoning and in many countries translation is effective from other languages word by word. Of course, induction and deduction models are generating mechanisms of relationships between inputs and outputs.

Deductive reasoning works from the "general" to the "specific", which is also called a "top-down" approach. It works as thinking of a theory about topic and then narrowing it down to specific hypothesis (hypothesis that we test or can test).

Civil Engineering	Mechanical Engineering	Chemical Engineering	Mining Engineering	.	Social Sciences
				.	

Fig. 2.3 Confined and leaky curriculum

Further down narrowing, if one would like to collect observations for hypothesis (note that one collects observations to accept or reject hypothesis and the reason one does that is to confirm or refute the original theory). In a conclusion, when one uses deduction one then reasons from general principles to specific cases, as in application of a mathematical theorem to a particular problem or in citing a law or physics to predict the outcome of an experiment.

In a valid deductive argument, all of the content of the conclusion is present, at least implicitly, in the premises. If the premises are true, the conclusion must be true. Valid deduction is necessarily truth preserving. If new premises are added to a valid deductive argument (and none of its premises are changed or deleted) the argument remains valid. Deductive validity is an all-or-nothing matter; validity does not come in degrees. An argument is totally valid, or it is invalid.

Even though Ph.D. means Philosophy of Doctorate implying that Ph.D. holder should know the basic philosophical structure about his/her research topics prior to any methodological applications, unfortunately in many research institutions all over the world, the students are trained in a rather dogmatic and mechanical manner. Hence, the end goal becomes in a dire "publish or perish" ideology without any slight, let along improvement or modification of the existing methodologies but only their mechanical uses in many applications through computers. For instance, Artificial Neural Networks (ANNs) are means of matching the given input data set to output data without knowing logically what is taking place between successions of layers. Unfortunately, classical education systems are based on very systematic, crisp and organized framework based on more than 23 centuries old Aristotelian logic, which has only two alternatives like black and white. Real life and its reflection as science have almost in every corner of information source gray fore and backgrounds. It is a big dilemma how to deal with gray information sources in order to arrive at scientific conclusions with crisp and deterministic logical principles. However, fuzzy logic principles with linguistically valid propositions and rather vague categorization provide a sound ground for the phenomenon concerned. The preliminary step is a genuine logical and uncertain conceptualization of the phenomenon with its causal and result variables that are combined through the fuzzy logical propositions (Fig. 2.4).

Such an approach helps not only to visualize the relationships between different variables logically, but furnishes a philosophical background about the mechanism of the phenomenon that can be presented to anybody linguistically without mathematical treatment. It is emphasized in Chap. 6 that in an innovative education system, the basic philosophy and fuzzy logic justifications in problem solving should be given linguistically prior to any crisp basis such as mathematics or systematic algorithms. In this way, the student will be able to develop his/her creative and analytical thinking capabilities with the support of teachers who are also trained or

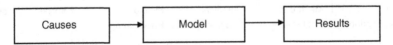

Fig. 2.4 Simple model of thinking

at least worked along similar directions. Since, the modern philosophy of science insists on the falsification of current scientific results, there are always room for ambiguity, vagueness, imprecision and fuzziness in any scientific research activity. Innovative education systems should lean more towards the basic scientific philosophy of the problem solving with fuzzy logical principles. However, many publications in recent years are software applications without grasps of basic principles based on linguistic and logical foundations.

2.9.1 Deduction

After general conceptions and principles movement of reasoning towards more specific directions to deduce a logically meaningful resuly is the trend of deduction. Read It works from the more general to the more specific scale, which is also occasionally referred to as a "top-down" logical approach. In the sequence of stages are first the proposal of a theory about the event, then its shrinkage into a narrower range through more specific *hypotheses* that can be tested; subsequently observations are collected to test the hypothesis with ultimate goal of a *confirmation* or rejection of the proposed theory (Fig. 2.5).

These four steps are in harmony with each other for the scientific explanation of the phenomenon concerned. Deductive reasoning starts usually with a theory followed by a hypothesis testing, which leads to observations for making the final decision as either to confirm or disconfirm the hypothesis and theory. This type of reasoning is based on crisp logic with two opposite conclusive alternatives. The evidences at hand may support theory after logical judgments such that observations or experimental results support, disconfirm, or irrelevant to a given hypothesis. The role of scientists is also justification that given all the available evidence, whether a hypothesis may be accepted as correct or approximately correct; may be rejected as false or both. Deductive inferences may help to assess evidential judgments. In the great majority of cases, however, the connection between evidence and hypothesis is non-demonstrative or inductive. In particular, this is so whenever a general hypothesis is inferred to be correct on the basis of the available data, since the truth of the data will not deductively entail the truth of the hypothesis. It always remains possible that the hypothesis is false even though the data are correct. One of the central aims in the philosophy of science is to give a principled account of these judgments and inferences connecting evidence to theory. In the deductive case, this project is well-advanced, thanks to a productive stream of research into the structure of deductive argument that stretches back to antiquity. The same cannot be said for inductive inferences. Although some of the central problems were presented incisively by Hume in the eighteenth century, our current understanding of inductive reasoning

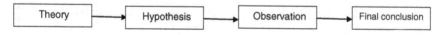

Fig. 2.5 Deductive reasoning

remains remarkably poor, in spite of the intense efforts of numerous epistemologists and philosophers of science [12].

In engineering even deductive reasoning is not used properly for logical final solutions. Unfortunately, most often classical engineering training implants in the minds of engineers that whatever they have learnt during education as formulations, equations, algorithms and software usages cover the wholeness of the problem and then engineer applies them directly to reach at the final calculations. Some may think that such a path is a deductive way of problem solution, but it is not, because although the four elements as in Fig. 2.5 are embedded in the engineering methodology, they are not considered specifically by engineers. Such engineering solutions cannot be considered as deductions, because there are no logical feelings by classically trained engineers. In order to feel the deduction way of reasoning, engineers should reason the logical principles (crisp logical rules) in their approaches in problem solving.

2.9.2 Induction

This type of inferential approach has more detailed way by moving from observations towards broader generalizations and theories. It is occasionally called as a "bottom up" approach and in this reasoning and inference system one begins to investigate the concerned event with specific observations and measures, and then tries to determine possible patterns and regularities. Subsequently s/he tries to formulate some tentative hypotheses, and finally, ends up with some general conclusions in the form of a theory (Fig. 2.6).

Inductive reason starts from specific observations or measurements and looks for possible patterns, classes, sets, regularities, hypothesis formulations that one could work with, and finally, moves towards general theory developments with plausible conclusions. The induction leads one to observe a number of specific instances and from which s/he can infer a general principle or a valid law. This reasoning is open-ended and more exploratory at the beginning.

Induction is implicative with conclusions that may have argument, which may go beyond the content of its premises. A correct inductive argument may have true premises and a false conclusion. New premises may completely undermine a strong inductive argument. Inductive arguments come in different degrees of strength. For creative engineering works this track of reasoning helps to arrive at innovative conclusions, devices and designs. Although the curriculum of engineering education is based on inductive pattern of teaching, but somehow most often the generation mechanism is not in accordance with this trend. The main reason is that there is not a logical sequence and interactive connections between the course contents or the way that they are thought in many engineering institutions.

Fig. 2.6 Inductive reasoning

The model of the inference of the best explanation is designed to give a partial account of many inductive inferences, both in science and in ordinary life. One version of the model was developed under the name 'abduction' by Pierce [15] (early in this century and the model has been considerably developed and discussed over the last 25 years). Its governing idea is that explanatory considerations are a guide to inference that scientists infer from the available evidence to the hypothesis which would, if correct, best explain that evidence.

According to inference to the best explanation, hypotheses are supported by the very observations they are supposed to explain. Moreover, on this model, the observations support the hypothesis precisely because it would explain them. Inference to the BE thus partially inverts an otherwise natural view of the relationship between inference and explanation. According to that natural view, inference is prior to explanation. First the scientist must decide which hypotheses to accept; then, when called upon to explain some observation, s/he will draw from her/his pool of accepted hypotheses. According to IBE, by contrast, it is only by asking how well various hypotheses would explain the available evidence that s/he can determine which hypotheses merit acceptance. In this sense, IBE has it that explanation is prior to inference [16].

The difficulties of the descriptive problem are sometimes underrated, because it is supposed that inductive reasoning follows a simple pattern of extrapolation, with 'More of the Same' as its fundamental principle. Thus one predicts that the sun will rise tomorrow because it has risen every day in the past, or that all ravens are black because all observed ravens are black. This model of 'enumerative induction' has, however, been shown to be strikingly inadequate as an account of inference in science. On the one hand, a series of formal arguments, most notably the so-called raven paradox and the new riddle of induction, have shown that the enumerative model is wildly over-permissive, treating virtually any observation as if it were evidence for any hypothesis. On the other hand, the model is also much too restrictive to account for most scientific inferences. Scientific hypotheses typically appeal to entities and processes not mentioned in the evidence that supports them and often themselves unobservable and not merely unobserved, so the principle of more of the same does not apply. For example, while the enumerative model might account for the inference that a scientist makes from the observation that the light from one star is red shifted to the conclusion that the light from another star will be red-shifted as well, it will not account for the inference from observed red-shift to unobserved recession [14].

Accordingto Hume, to justify induction we would have to produce a cogent argument whose conclusion is that induction is generally reliable and whose premises are not themselves inductively based. The only such premises are reports of past observation and the demonstrative truths of logic and mathematics. All cogent arguments are either deductive or inductive. Now we face a dilemma. There can be no cogent deductive argument for the reliability of induction, since no number of past observations (along with demonstrative truths) deductively guarantees that induction is generally reliable. In particular, past observations will never entail that induction will be reliable in the future. Neither is there a cogent inductive argument for induction, since any such argument presupposes the very practice it is supposed to justify. For example, to argue that induction is likely to be reliable in future on the grounds that it has been reliable in the past would beg the question, even if it were granted that the past reliability of induction could itself be known on the basis of observation.

The constructive empiricist is no inductive skeptic, since to say that all the observable consequences of a theory are true is a much stronger claim than to say merely that its observed consequences are true; but the realist goes further by sanctioning in addition vertical inferences to the truth of a theory's claims about unobservable entities and processes. Perhaps the best known example of this application of inference to the best explanation in defense of scientific realism is the so-called 'miracle argument', discussed by Putman [17]. He takes it that the model provides a good solution to the descriptive problem and proposes that philosophers may themselves make an inference to the best explanation in defense of scientific realism. Suppose that all the many and varied predictions derived from a particular scientific theory are found to be correct: what is the best explanation of this predictive success? According to Putnam [17], the best explanation is that the theory itself is true. If the theory were true, then the truth of its deductive consequences would follow as a matter of course; but if the hypothesis were false, it would be a 'miracle' that all its observed consequences were found to be correct. So, by a philosophical application of inference to the best explanation, we are entitled to infer that the theory is true, since the 'truth-explanation' is the best explanation of the theory's predictive success. This higher-level inference is supposed to be distinct from the first-order inferences scientists make, but of the same form. This justificatory application of inference to the best explanation has considerable intuitive appeal, but it faces three objections. The first is that the truth-explanation for the predictive success of a theory is not really distinct from the substantive scientific explanations that the theory provides and on the basis of which it was inferred by scientists in the first place. If this is so, then the miracle argument provides no additional reason to believe that the hypothesis is correct: it is merely a repetition of the scientific inference it was supposed to justify. This objection can be answered, however, by observing that the two sorts of explanation have a different structure. The scientific explanations a theory provides are typically causal, whereas the truth-explanation is logical. The truth of a theory does not physically cause its consequences to be true; the explanatory connection is rather that a valid argument with true premises must also have a true conclusion. The second objection to the miracle argument is that, even if the truth explanation is distinct from the scientific explanations, the inference to the truth of the theory is vitiated by the same sort of circularity that Hume appealed to in his skeptical argument.

2.9.3 Analogy

It is an inference from one particular case to another, as opposed to deduction and induction, where at least one of the premises or the conclusion is general. The word analogy can also refer to the relation between the source and the target themselves, which is often, though not necessarily, a similarity, as in the biological notion of analogy (http://en.wikipedia.org/wiki/Analogy). For instance, Niels Bohr's model of the atom made an analogy between the atom and the solar system (Fig. 2.7).

Fig. 2.7 Analogy reasoning
(http://en.wikipedia.org/wiki/
Analogy)

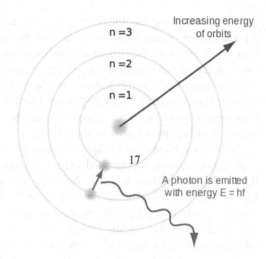

There are many phenomena that can be explained linguistically almost with the same sentences but on the basis of different terminology. The simplest analogy between the scientific laws concentrates into a single sentence as follows: *Two variables are directly and linearly proportional with each other.*

This sentence covers many similarities in different scientific disciplines. For instance, if the two variables are force and acceleration then the Newton's law can be understood from the sentence. In case of groundwater velocity and hydraulic gradient the law is Darcy's suggestion. Another example to analogy can be found between heat transfer and groundwater movement in aquifers [1].

2.9.4 Intuition

Intuition is another type of reasoning, which is used by most frequently by youngsters and to a lesser extent by elders. It may appear instantaneously in one's mind after seeing or looking at something or some thought and as a result of reflection one can make quite rational and logical guess. Although it may be rather rudimentary, but very useful in giving a starting point from which induction or deduction can follow. It is the chief type of reasoning used by early elementary students, and students must be shown the flaws in it by the use of cognitive conflict in order to learn to move past intuition towards induction and deduction.

2.10 Approximate Reasoning

Reasoning is the most important human brain operation that leads to creative ideas, methodologies, algorithms and conclusions in addition to a continuous process of research and development. Reasoning stage can be reached provided that there is stimulus for the initial driving of mental forces. Ignition of pondering on a phenomenon comes with the physical or mental effects that control an event of concern. These effects trust imaginations about the event and initial geometrical sketches of the imaginations by simple geometries or pieces and connections between them [21]. In this manner, the ideas become to crystallize and they are conveyed linguistically by means of a native language to other individuals to get their criticisms, comments, suggestions and support for the improvement of the mental thinking and scientific achievement.

Approximate reasoning helps to resurface in information technology, where it provides decision support and expert systems with powerful reasoning bound by a minimum of rules and it is the most obvious implementation for the fuzzy logic in the field of artificial intelligence (Chap. 4). It is already explained how one can easily relate logic to ambiguous linguistics in forms of different fuzzy words such as 'very', 'small', 'high', and so on. Such flexibility allows for rapid advancements and easier implementation of projects in the field of natural language recognition. Fuzzy logic brings not only logic closer to natural language, but closer to human or natural reasoning. Many times knowledge engineers have to deal with vague and common sense descriptions of the reasoning leading to a desired solution. The power of approximate reasoning is to perform reasonable and meaningful operations on concepts that cannot be easily codified using a classical approach. Implementing the fuzzy logic will not only make the knowledge systems more user friendly, but it also will allow programs to justify better results. Detailed information is given about the fuzzy logic in Chap. 4.

The qualitative jump of consciousness to a higher level results in transcending the fuzziness. As far as consciousness is of a holistic characteristic of human, and perhaps not just of human, but also nature and not only a product of mind, but its growth and transformation are possible when the factors responsible for the integrity of all three inseparable constituents of human individuality, which are body, mind and soul, become simultaneously activated. This simultaneous activation ('firing' or 'triggering') is referred to as a consciousness resonance and hence: *The fuzziness of understanding can be transcended when the consciousness resonance occurs.*

The consciousness resonance is a resonance of all factors responsible for human integrity as manifested in the holistic nature of consciousness. What are these factors? First, factors, which contribute in keeping human body healthy and human mind capable to think and decide, no matter what kind of logic it prefers—fuzzy, binary, inductive, deductive, abdicative, etc. However, these factors are not enough. The consciousness resonance cannot occur when neglecting the soul factors; one can name some of them as sensitivity and responsiveness, awareness and ability to stay awake, passionate desire to get out of the 'attractor' of egocentric thoughts and desires, compassion and love, willingness to explore more subtle and spiritual dimensions of reality and to share with others skill, knowledge and wisdom [4].

The consciousness resonance does not eliminate fuzziness, which is an eternal companion to any process of thinking and knowing. At the same time, when the consciousness resonance helps one to transcend the fuzziness related to a problem that dissolves, it opens space for new problems to emerge bringing with them new fuzziness to puzzle our thoughts and feelings. At any level of consciousness, there are infinite number of phenomena and processes challenging the 'swarm' of our perceptions, of our beliefs and hopes, views and attitudes, aspirations and dreams.

The reasoning (philosophy of fuzzy thinking) is based on graded concepts. It is a concept in which everything is a matter of degree, i.e., everything has softness (elasticity). The fuzzy logic theory has been given first in its present form through the early publications of Lofty Asker Zadeh [24]. He wanted to generalize the traditional notion of a set and a statement to allow the grades of memberships and truth values, respectively. These efforts are attributed to the complications that arise during physical modeling of real world. These are,

- Real situations are not crisp and resolute; hence they cannot be described precisely;
- The complete description of a real system often would require by far more detailed data than a human being could ever recognize simultaneously, process and comprehend.

The last statement Zadeh calls the *principle of incompatibility*. Its message is that the closer one looks at a real-world problem, the fuzzier becomes its solution.

The subjectivity, i.e., dependence on personal thoughts is the greatest at the perception stage and as one enters the visualization domain, the subjectivities decrease and at the final stage since the ideas are exposed to other individuals, the objectivity becomes at least logical, but still there remains some uncertainty (vagueness, incompleteness, missing information, etc.), and hence, the final conclusion is not crisp but fuzzy. Fuzzy reasoning always exists in scientific domains, but in the classical and mechanical approaches they are deleted artificially by idealizations, isolations, simplifications and assumptions.

The classical logic renders the final stage in solutions into crisp forms by defuzzification, which means neglecting all the uncertainties either through the assumptions or through a safety factor or confidence interval in many engineering solutions. Crisp reasoning conclusions do not provide soft domain for further research especially in many aspects of engineering. Therefore, classical methodologies and formulations are fragile, hard and difficult to accept the consequences. In order to avoid the crispness, the statistics and axiomatic probability concepts are suggested, but they are also based on the classical logic, where the consequences are black and white without gray tones, which is available in approximate reasoning through fuzzy logic principles and modeling.

Finally, all the conclusions must be expressed in a language, which can then be converted into universally used symbolic logic based on the principles of mathematics, statistics or probability statements. This explanation shows that fuzzy logic is followed by symbolic logic (mathematics). Unfortunately, in many educational systems all over the world, this sequence of language and symbolism is overturned into the sequence of first symbolism (mathematics) and then linguistic understanding which is against the natural reasoning abilities of human. This is especially true for countries or societies who are trained with symbols and those when they return to their community, the first difficulty is to convey the scientific messages in his/her language, and therefore, in order to avoid such a dilemma the teacher bases the explanation on symbolic logic. This is one of the main reasons why scientific thinking and reasoning are missing in many engineering institutions all over the world. The avoidance of such a problem is approximate reasoning where the facts are explained through natural languages first.

2.11 Criticism (Dialectic)

If engineers depend without reasoning and critical review on the methodologies and formulations then they may accept derivations or suggestions of these methods rather blindly, which may lead to various dead-end in problem solving. If critical review and logical bases of these methodologies are not questioned, then engineers and engineering chambers may not be in mutual support on each other. In many countries engineering chambers are under the influence of politics, and therefore, cannot generate new and innovative ideas, and consequently, neither philosophy of engineering nor logical inferences can take place in such media for creative idea generations. Any engineering methodology must be criticized according to the environmental conditions rather than their blind applications throughout years.

In order to have a critical view, engineers must suspect from the proposed solutions and try to minimize the harmful situations, but maximize the benefit from the application of each methodology. Perhaps, the best solution is a mixture between these two stages as hybrid methodology leading to optimum solutions. Engineering phenomenon may not have strictly abstractable logical and mathematical principles and in each case due to involvement of a material, the solutions are not rational only but their verifications need experimental bases also. Content of an engineering work may not be as the content of a scientific study. In scientific studies there are also materials and their variability constitutes scientific features, but in engineering works, materials are given forms and geometrical shapes, and therefore, their behaviors are investigated separately by using scientific and philosophical principles under the light of logical steps. Similar to scientific evolution as a dynamic process, engineering tasks should also be dynamic leading to new developments. Unfortunately, today most often engineering procedures are considered as master keys to many problem solutions through direct application of already obtained scientific findings. Such approaches narrow the development ways of engineering without complete innovative methodological suggestions.

In order to show this point, let us consider Newton's second law, where the force is directly and linearly proportional with acceleration provided that the mass is constant. The following points are among the possible critical questions that may shed some doubts about the general validity of this law.

- Why there is a linear relationship?
- Why mass is assumed constant?
- Why directly proportional relationship exists?
- Is this law valid at any time and space?

Fig. 2.8 Force-acceleration relationships

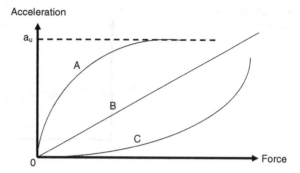

Search for proper answers to each one of these questions requires philosophical thinking principles. Philosophical and logical reasoning on these points including methodological, mathematical, algorithmic approaches and theories, equations, formulations and laws are already accepted as default as always valid and correct. Although one cannot reach to absolutely correct results, but at least such questionings provide a way of avoiding the memorization and direct blind application of them in engineering. In order to answer to first question, one can think just the opposite situation. The relationship (human can think about two variables' mutual relationship at a time) is equivalent to the association degree between two variables and this statement can be shown geometrically as one relationship with three alternatives (A, B, and C) on a Cartesian coordinate system (see Fig. 2.8).

There are two non-linear and one linear possibility for directly proportionality. The most logical and rational one among these three alternatives can be selected by reasoning from various points of view. Prior to such a selection, it is useful to start with the concepts of each variable critically. Force is a multiplicative variable but acceleration is the change of velocity by time, or even the change of change of distance by time. Let us first consider case A linguistically and criticize it on the basis of rational reasoning. This alternative implies that as the acceleration increases force also increases, but for large acceleration values its increase reaches to an almost stable level. One can deduce that acceleration may reach to an upper level as a_u after which force increases continuously in a manner that does not show a balanced proportionality. This is not a plausible conclusion, and hence, case A is illogical and it can be eliminated. With similar arguments the reader can judge that case B is also not plausible, and hence, there remains only C as rationally acceptable. After all what have been explained, now any scientist or engineer has the bases of $F = ma$ formulation philosophically and logically rather than its static existence in the memory for ready applications. Without above simple reasoning the static case in the memory can be transferred to other individuals in the society, and hence, instead of mind experiment, rational judgment and criticism memorization prevails.

Why the mass is constant? In fact, mass changes always in our daily life. A car or an airplane changes its mass as it moves by fuel consumption. This means that Newton's second law can be used only instantaneously. One can visualize that mass decreases in an exponential form as in Fig. 2.9.

Fig. 2.9 Mass-time variations

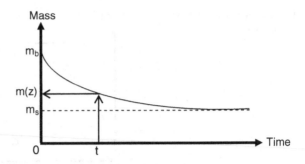

It is necessary to measure the mass at each time instant and then enter this value in the basic law for force calculation provided that the acceleration is known simultaneously. One can represent the shape in Fig. 2.9 by a decreasing exponential function, which has, in general, the following form.

$$m(t) = m_s + (m_b - m_s)e^{-c_m t}$$

where c_m is a mass constant and the substitution of this expression into the Newton's second law leads to,

$$F(t) = \left[m_s + (m_b - m_s)e^{-c_m t} \right] a$$

In this manner, the constancy of mass has been avoided according to its exponential decrease by time. However, velocity and hence acceleration also vary as time passes, and therefore, their variations may also be represented by some functions. Let us consider that the velocity increases as in Fig. 2.10 which reaches a constant value after some time.

By definition acceleration is the change of velocity by time; therefore, visually one can derive from this figure the change of acceleration by time as another exponentially decreasing function given in Fig. 2.11.

The mathematical form of acceleration variation can be written as a decreasing exponential function,

$$a(t) = a_b e^{-c_a t}$$

Fig. 2.10 Velocity-time variations

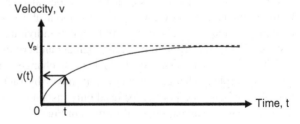

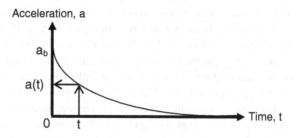

Fig. 2.11 Acceleration-time variations

where c_a is the acceleration constant. Substitution of this equation into the force expression leads to the following equation, which perhaps is the first time the reader comes across.

$$F(t) = \left[m_s + (m_b - m_s)\, e^{-c_m t}\right] a_b e^{-c_a t}$$

It is obtained after the philosophical thinking as explained above by allocating negative exponential forms to mass and acceleration variations. However, many other variation forms may lead to different alternatives of the force variation, and therefore, a scientist or engineer with the support of the philosophy of engineering could develop similarly any case that is suitable for his/her thoughts. The last force equation is renewable under the prevailing conditions by considering philosophical (verbal reasoning) and logical rules. This example shows that like new technologies and productions even the basic equations can be enriched according to prevailing conditions and new formulations can be obtained for practical applications. If one avoids philosophy of engineering and logical reasoning principles then for him/her the only remaining solution will be the formulation that has been learned the first time as it is in the memory without reasoning, and consequently, such an engineer cannot adapt his/her thoughts to new situations. Innovative thoughts cannot be without philosophy or rational thinking and reasoning. Philosophical solutions are not the final destinations, since philosophy does not have a finite limit. However, for anything that is questionable its solution can first be achieved philosophically leading to the generation of many related ideas and further problems. Hence, the door of philosophy is open always for those who would like to enter into rational, logical and critical thought experiments. Engineering necessitates practical applications of the philosophical products as formulations. It is necessary that an engineer should be acquaintant at least with basic principles of philosophy for innovative inventions and consequent applications.

Another suspicion is whether Newton's second law is valid at any place any time? The reader can provide answers with critical thinking. Initially the validity of this law was respected at any place and time. Later, it is understood that it is valid for medium scales, whereas at quantum or at very big scales (relativity theory), it is not valid. In the quantum case, it is not possible to identify cause-effect relationship deterministically but probabilistically or statistically.

For instance, the main philosophical issue in the interpretation of Einstein's equation is whether mass and energy are the same property of physical systems and whether there is any physical sense in the conversion of mass into energy or vice versa. There are different interpretations and arguments on this equation by different authors [2, 7, 8, 9]. All the discussions have revolved in the frame-work of the special relativity, where the light velocity constancy is the fundamental assumption. However in the following, first of all, the simple derivation of energy is given in the Newtonian domain and then its philosophical extension to Einstein's equation is presented. Hence, interpretations of energy in velocity domains less than the light velocity may shed some light on the discussion on whether the mass and energy are converted to each other. First, the force, F, is defined as the change of momentum in the Newtonian physics as,

$$F = \frac{d(mv)}{dt} \tag{2.1}$$

where mv is the momentum, m is the mass and v is the velocity of this mass. It is well-known that absorbed radiation of energy is accompanied (in classical theory, in quantum theory, and in experiment) by a radiation momentum is the ratio of energy, E, to light velocity, c as $E/c = mc$ [23]. For example, from quantum theory $E = h\upsilon$ and $\lambda = h/p$. Since, $\lambda\upsilon = c$, this gives $(h/p)(E/h) = c$ or $p = E/c$. In the derivation of Einstein's equation, conservation of momentum is used and likewise herein the change of momentum is taken as a basic approach to this problem first simply in the Newtonian physics domain. Energy, E, which is equivalent to work, FdL where dL is the distance covered. Hence, Eq. (2.1) can be rewritten as,

Since basic definition of velocity is $v = dL/dt$, then this last expression takes the following form,

$$dE = d(mv)v$$

This expression can be expanded as

$$dE = v^2dm + mvdv \tag{2.2}$$

which indicates that energy variation consists of two gradients, namely, change in mass (the first term on the right hand side, related to Einstein's energy) and change in velocity (the second term on the right hand side, related to Newtonian kinetic energy). Mass and energy are regarded as distinct properties because in Newtonian physics as in Eq. (2.2) they are distinct and measured in different units. This is due to the fact that spatial and temporal units are perceived separately, which gives rise to the perception of different mass and time properties. The expression in Eq. (2.2) can be expressed verbally as, *Energy change is the summation of a constant (velocity) times change in mass plus another constant (mass) times velocity times change in velocity.* This expression includes both large scale (relativity) and ordinary scale (common sense) perceptions. To this end the following two questions can be asked as what is the energy, E_c, if the velocity (light velocity) is constant? This case corresponds to the assumption of special relativity where the velocity is assumed constant as light velocity, c, only [5]. The answer is that the second term in Eq. (2.2) becomes zero because there is no velocity change and hence the energy remains as,

$$dE_c = c^2 dm$$

This gives the total energy, E_{ct}, due to mass change after integration as,

$$E_{tc} = mc^2 \qquad (2.3)$$

This is the most well-known Einstein's energy equation in the special theory of relativity. According to special relativity light travels at the same speed for all inertial observers, which implies that one can select units such that spatial distances are specified by units of time (space–time concept). In such units energy and mass have the same units and they are equal numerically, which implies that mass and energy are not two distinct properties. In a way the perception of mass and energy as distinct units is due to the fact that spatiotemporal intervals are overlooked.

The second question is what is the energy, E_m, if the mass is constant? In this case, the first term on the right hand side of Eq. (2.2) is zero and the energy expression takes the form as,

$$dE_m = mvdv$$

which after the integration yields the total energy due to the velocity change only as,

$$E_{tm} = \frac{1}{2}mv^2 \qquad (2.4)$$

This is the kinetic energy expression in the Newtonian physics domain. After all these simple derivations from Eq. (2.2) indicates that mass and energy are not the same as suggested by some philosophers and physicists. The same equation implies that mass and energy are distinct properties of physical systems.

2.12 Pragmatism

One of the thinking systems in recent years is pragmatism. It is a Greek word that means work of affair, and hence, continuity, dynamism and benefit. The life is full of work, affairs, transections and activity. In order to direct these works and affairs one needs to think and by philosophical thoughts to direct them towards the right targets. Although the basis of philosophy is to search for absolute rightness, in the pragmatic philosophy the rightness varies depending on the purpose. Thus, in the pragmatic thoughts rightness has a multitude of dimensions and it is a relative affair. It is not a dependable way of thinking in engineering, but it provides many alternatives among which there is one that suits the present engineering conditions with solution. However, there are also dangers in this thought system. For example, an engineer cannot make cheats in design or construction stages, say, by using cheap material so as to have cheaper solution for the same plan and project.

Such a behavior is not ethical at all. Pragmatism is a way to determine rightness according to supporters or client requests for making him/her happy at the cost of some engineering principles. However, pragmatism is in existence in many affairs all over the world mostly for higher returns economically.

2.13 Rationality

Many people, although, have heard the word "rationality" do not know or cannot explain its meaning fully. One may insist that all the scientific knowledge is rational. Even some go further to the level of faith in their allegation. One should not condemn them because classical education system might train them in these ways. The etymologic meaning of each word leads one through critical questioning (why?) towards more detailed epistemological knowledge content. It is not enough to know only but one should also perceive what the knowledge foundations are. For rational thinking one must imagine the objects and then visualize it through designs and finally produce information. For example, around us there are points, lines, planes, triangles, etc. Each one of these is an element that is perceived by our minds. These are ideal phenomena. Based on such basic shapes one can constitute his/her conceptions, visualizations, and finally, knowledge production. Thus, among the basic elements of any engineer, geometry enters the view. These basic shapes when combined together in a meaningful manner lead to knowledge production in the mind and they also help to trigger further ideas about the phenomenon of concern. Thus many related ideas can accumulate in the mind. Let us consider a triangular shape as in Fig. 2.12.

This is the simplest shape that an engineer uses in many designs. In order to benefit from this basic shape first of all engineer should look at it from different angles with rational thinking under philosophical questions for further exploration about its properties. Accumulation and ripeness of these information and knowledge about the triangle help to further systemize scientific solutions with practical engineering applications. In this manner engineer by taking the flavor of knowledge production can stimulate himself/herself for further productions and in a way encourages himself/herself. Thus trains and educates his/her mind. In any future problematic situation these bases and product information and knowledge help engineer to design better structures and programs. Always critical questioning and rational reasoning must not be forgotten with philosophy and logical principles. One can increase

Fig. 2.12 Triangular shape

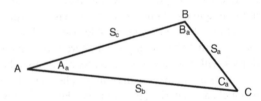

his/her knowledge level by criticizing the available knowledge. After following this path, there is not even need for making experiments, because with rational thinking an engineer can reach crisp and un-doubtful knowledge. Similar to naming a newly born baby, knowledge production also requires naming new products by words, i.e., names. This helps to provide knowledge transmission easily, and afterwards, it will be mentioned and remembered with the same name between the specialists.

A first glance to Fig. 2.12 shows that any triangle has three sides and their intersections are named as "corners". Accordingly, it is necessary to name each corner separately and let them be denoted by A, B and C. Each corner is generated by the intersection of two sides or more generally it is a vertex on a broken line. One can also realize that there is an angle between the two sides. These can be labeled as angles A_a, B_a and C_a. On the other hand, each line opposite to each vertex can also be labeled, say, as S_a, S_b and S_c. All these labeling indicate that even a single triangle has many elements that are different from each other.

After the imagination and visualization of all these elements in mind, now it is time to critically question "what are the other linguistic and numerical features about all these features as well as the relationships between them". Let us first consider to find possible relationships between the same types of elements. For this purpose, in Fig. 2.13 a parallel line is drawn to side S_b and then the relationships among the angles are sought.

A very close rational inspection indicates that at corner B summation of all internal angles is equal to a "straight line angle", which is a very important conclusion. This is rule of "straight line angle" that is valid irrespective of triangle shape with absolute accuracy. A sub-inference of this result is that any straight line angle is equal to two right angles (see Fig. 2.14). Finally, one can conclude by saying that the summation of internal angles in any triangle is equal to two perpendicular angles.

If the question is what is the right angle?, then the answer is that in such an angle projection of any side on the other side is equal to zero.

Any reader with classical educational training will jump to the conclusion that right angle is equal to 90° and straight line (two right angles) angle is 180°. These numerical are not anything than the mind has attached for convenience. Let the reader think whether the summation of internal angles of a triangle is equal to 180°, or the same summation is linguistically equal to two right angles'

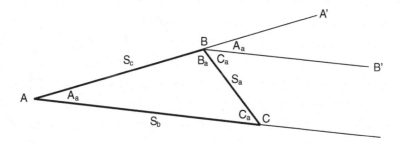

Fig. 2.13 Triangle angle relations

Fig. 2.14 Right angle
definitions

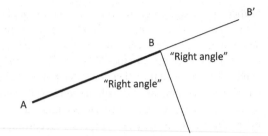

summation is more rational, logical and general. The linguistic inference is the reality but instead of 180° one can also adapt 200°, which does not affect the linguistic case. This shows that rather than numerical labeling, linguistic inferences are correct and they have general knowledge contents. All these derivations are based on the geometric design concepts, which are main tools in any engineering work.

Let us now think about some other questions, say, is there any relationship between the sides? Rational reasoning dictates after a slight thought that in any triangle "summation of any two sides is greater than the third side". Otherwise one cannot obtain a triangle.

All what have been explained above about the triangle are valid in case of plane geometry. If the triangle is on a spherical surface then spherical geometry and different relationships become valid. An example for this case is given in Fig. 2.15 as ABC triangle.

It is obvious that the triangle on the northern hemisphere has two right angles on the equator and thus the summation of internal angles cannot equal 180°, in fact the summation is more than this.

If one asks whether there are benefits from triangles in engineering? Then it becomes obvious that they exist in many engineering design. Especially, in field topographic surveys the surface features of any area on the earth can be visualized through a set of triangles, which is named as "triangularization". It is possible to draw elevation contour lines after measuring through surveying instruments with

Fig. 2.15 Spherical triangle

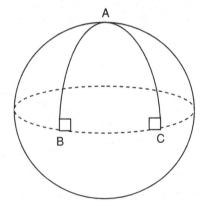

Fig. 2.16 Truss system

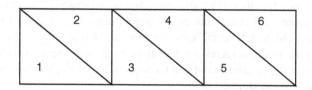

the elevations and coordinates of each corner of triangular in the triangulation network. By means of each triangle one can know the slope of the surface locally. On the other hand, in civil engineering, triangles are put together, and hence, truss systems are produced for load carriages as in Fig. 2.16.

The triangle with number 1 preserves its shape even without welding at the corners but with rivets only. All other triangles in this truss bare exactly the same property with the first one and with others (2, 4 and 6), which are turned 180°. Addition of similar triangles may enlarge the scale of the truss according to the desire of engineer.

2.14 Experimentation

Perception of knowledge only by rational reasoning is not enough and in order to confirm it, experimentation is necessary. One may stick to the idea that rational inferences may although be based on logical bases; their final confirmation should be achieved through convenient test experimentally. Experimentation can be regarded as knowledge engine and rational thinking may cause errors or even wrong inferences by mind experiments. The numerical determination of the elasticity modulus necessitates performance of a set of experimental study. For this purpose, samples of certain shapes from the same material, say sticks made of steel are subjected to tensile strength and the deformation is measured as the stress increases, and hence, a scatter diagram of the measurement set is obtained (see Fig. 2.17). The reason of scatter around a straight-line may be due to the material heterogeneity, measurement errors, irregularity in the geometric shape,

Fig. 2.17 Experimental elasticity modulus determinations

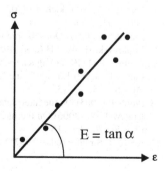

$E = \tan \alpha$

etc. or altogether. If ideally the samples and materials have perfect homogeneous, isotropic and cylindrical cross-section then all the measuring points are expected rationally to fall on the same straight line. Even without any further methodological use (such as regression technique in the statistics) one can fit the best straight-line with eye adjustment and the slope of this line gives the numerical value of the elasticity modulus.

2.15 Engineering and Skepticism

Whatever is the engineering design, it is better to review the plan or project with skepticism, or to invite some independent reviewers to look at the issue from their point of view. This approach may bring additional views from different angles and engineer may be empowered with improvements in the original and uncountable ideas. This gives to engineer opportunity to defend his/her opinion from different skepticism. In such a defense the engineer should know much of epistemological knowledge for the case so as to stand firmly and at times elastically against the oppositions. Skepticism opens ways to various knowledge types with supportive improvement in defense. For instance, most of the empirical equations in engineering must be viewed from skeptical view, because many of them are proposed under a set of certain circumstances, which may not be valid in the case of current engineering tasks. Skepticism sharpens rational thinking channels with consensus and common agreement between experts. Especially expert views are subject to skepticism to some extent; otherwise, if they all match the same template then they cannot be regarded as expert view. In general, a priori knowledge is not attacked by skeptic ideas compared to empirical knowledge.

References

1. Bear J (1979) Hydraulics of groundwater. McGraw Hill Book Co., New York, p 567
2. Bondi H, Sprungin CB (1987) Energy has mass. Phys Bull 38:62–63
3. Dimitrov D (2000) On liberal collective identity. Notes on Intuitionistic Fuzzy Sets 6(2):42-51
4. Dimitrov V, Korotkich V (2002) Fuzzy logic: a framework for the new millennium. Physica-Verlag, Heidelberg
5. Einstein A (1905) Does the inertia of a body depend upon its energy content? In: Einstein A, Minkowski H, Weyl H (eds) (1952) The principles of relativity, pp 69–71 , New York
6. Edgington D (1997) Vagueness by degrees. In: Keefer and Smith (eds) (1999), pp 294–316
7. Flores F (1999) Einstein's theory of theories and types of theoretical explanation. Int Stud Phys Sci 13:123–133
8. Flores F (2005) On the interpretation of the equation $E = mc^2$. Int Stud Phys Sci 19:245–260
9. Krajewski W (2006) On the interpretation of the equation $E = mc^2$: reply to Flores. Int Stud Phys Sci 20:215–216
10. Keefe R, Smith P (eds) (1997) Vagueness: a reader. MIT Press, Cambridge

11. Lassiter D (2011) Vagueness as probabilistic linguistic knowledge. In: Nouwen R, van Rooij R, Sauerland U, Schmitz H-C (eds) Vagueness in communication. Springer, Berlin, pp 127–150

12. Lipton P (2000) Inference to the best explanation. In: Newton-Smith WH (ed) A companion to the philosophy of science. Blackwell, pp 184–193

13. Machina KF (1976) Truth, belief and vagueness. J Philos Logic 5:47–48

14. Newton-Smith WH (2002) A companion to the philosophy of science. Blackwell Publisher, Malden

15. Peirce CS (1931–1935) The collected papers of Charles S. Peirce, 8 vol. Harvard University Press, Cambridge

16. Psillos S, Curd M (2008) Philosophy. The Routledge Companion to Philosophy of Science. p 194

17. Putman H (1978) Meaning and the moral sciences. Routledge, London

18. Russell B (1948) Human knowledge: its scope and limits. George Allen and Unwin, London

19. Şen Z (2002) Bilimsel Düşünce ve Mühendislikte Matematik Modelleme İlkel, Su Vakfı

20. Şen Z (2009) Bulanık Mantık İlkeleri ve Modelleme (Mühendislik ve Sosyal Bilimler). Su Wakfı Yayınları. (Fuzzy logic principles and modeling—Engineering and Social Sciences) (in Turkish)

21. Şen Z (2010) Fuzzy logic and hydrological modeling, Taylor and Francis Group, CRC Press, p 340

22. Şen Z (2006) Batmayan Güneşlerimiz. (Our unsetable suns). Altın Burç Yayınları

23. Shadowitz A (1968) Special relativity. Courier Dover Publications, p 203

24. Zadeh LA (1965) Fuzzy sets. Inf Control 8(3)338–353

Chapter 3
Philosophy and Engineering

3.1 General

Even though the philosophical thoughts are as old as human history, it became systematized during the old Greek period starting from almost sixth century before Christ. During this era, the rational thinking is based on the philosophical inferences without training and testing, and therefore, they were all speculative thought generations. Although many of such rational statements were proved to be valid, later many others remained as invalid. Philosophy is the art of thinking freely without any restriction about any event or phenomenon. Early humans were stimulated towards thinking by getting feedbacks from objects of natural or artificial characters. Any branch of knowledge and information were gathered under the umbrella of philosophy without distinction between events.

The philosophical invitation into engineering aspects helps to explain the epistemological, methodological, ontological and ethical issues, out of which for the last 2–3 decades ethical issues only are welcomed in engineering institutions leaving aside other issues. However, inclusion of other issues in a convenient proportion will broaden the view horizons of engineers for better productions. These issues may be explained briefly as follows.

(1) Epistemological aspects of engineering: Here the questions are what is the nature of the engineering knowledge? And how are they justified? This also implies to what extent engineers should grasp knowledge and what are the types and information content of engineering know-how? Detailed examination of engineering knowledge may create new philosophical dimensions. Epistemologically scientific, technologic and engineering thinking patterns are not completely distinct but have significant differences leading to useable knowledge. From epistemological point of view "cogito ergo sum" ("I think, therefore I am") is the basis of philosophy [1].

(2) Methodological aspects of engineering: Any kind of engineering method should be questioned as its generation mechanism, functional elements, justification

Z. Şen, *Philosophical, Logical and Scientific Perspectives in Engineering*,
Intelligent Systems Reference Library 143, DOI: 10.1007/978-3-319-01742-6_3,
© Springer International Publishing Switzerland 2014

aspects and measurement indices take place. For instance, Koen [2] states that all methods in engineering are heuristic (based on human sense) in nature.

(3) Ontological aspects of engineering: This is concerned about the status of engineering functions and designs. If there are philosophical assumptions then they must be brought down to engineering understanding levels for relevancy. Engineers are concerned with applied ontology for their end product purposes.

(4) Ethical aspects of engineering: This is the mostly cared for philosophical issues in engineering as well as in many other disciplines for the last three decades. This is the area where engineers make contributions in addition to philosophers.

Concerning all these points, engineers and philosophers can work on conceptual aspects so as to come into common area with each other, even though there will remain disagreement between them to a certain extent.

3.2 Philosophy and Branches

Philosophy implies various general and fundamental problems that are rather vague and subjective such as existence, knowledge, values, reason, mind, and especially language, which need rational consensus among the philosophers. Likewise, although many researchers may not sense that their issues are full of philosophical arguments, they try to settle down the problems by a set of rational and logical rules linguistically prior to any other means. In any problematic issue critical and systematic arguments pave the way towards rational solutions through approximate reasoning, because in philosophy any issue does not have crisp and deterministic end. The ends of present time are always open to further arguments for better improvements. Different branches of the philosophy and science philosophy are not mutually inclusive, but have many common points and arguments, which indicate that there is always uncertainty and fuzziness in linguistic inferences. Metaphysics as one of the main branch of the philosophy need deep and expansive reasoning for fruitful inferences about the nature of reality, where different sorts of mutual relationships are sought internally and externally for meaningful and useful conclusions about any event or among events of any type including engineering aspects. The basis of relationship search is to try and relate causative (inputs) to desired result (output) factors. The preliminary triggering means in relationship search are truth, belief and justification, which are in themselves subjective, but common sense and consensus provide solution, which may be improved later by others. Truth, belief and justification cannot be achieved without etymology and epistemology, because these constitute the origin and meaning knowledge. Absolute belief and justification are not possible in the philosophy and science. This is the main reason why philosopher and scientist are skeptical in the relationships searches. Since scientists are skeptical about the results so should be engineers who use the end products of scientific researches.

As in many careers and also in engineering moral philosophy (ethics) started to emerge in the last 3–4 decades as more dominant than any other philosophical aspects. Ethical virtues give one the way of interactive and justifiable action in his/her daily career life. Unfortunately, today moral aspects are overwhelmingly emphasized as philosophical aspects, but especially in engineering reasoning, thinking and intuitive creation facets of innovative deals are not cared for better training or education. As for another engineering branch, which has been forgotten to a significant extent in our days, is the aesthetics, which deals with beauty, art, enjoyment, sensory-emotional values, perception, and matters of taste and sentiment. So, without such philosophical concepts engineers started to run after ready scientific formulations and template solutions for the problem at their hands. However, those who care for more uncertainty, fuzziness, aesthetics, and creative difference, from existing conventional solutions, are creative outliers among their career partners. Is it possible to achieve, say, software programming without the philosophy of the topic and a set of logical rules? Of course, the answer is no, because although there are many formulations, algorithms, procedures and heavy mathematical solutions, an engineer without philosophy and logic is incapable to alter, modify or completely suggest a new approach to the same problem.

Logic is the most extensively employed mind activity under the umbrella of philosophy, which even has almost transpassed the philosophical boundaries as for the practical applications are concerned. It is among the most intensive content of this book. One can also classify other branches as language and philosophy, religion and philosophy and for the last couple of year even engineering philosophy. As for the universities are concerned, the most cited type of philosophy is the philosophy of science, where philosophies of mathematics, physics, chemistry, genetics, anthropology, psychology and many others can be viewed. An alternative approach to philosophy of science and scientific knowledge has been explained by McMullin [3].

Quinton [4] stated that "Philosophy is rationally critical thinking, of a more or less systematic kind about the general nature of the world (metaphysics or theory of existence), the justification of belief (epistemology or theory of knowledge), and the conduct of life (ethics or theory of value). Each of the three elements in this list has a non-philosophical counterpart, from which it is distinguished by its explicitly rational and critical way of proceeding and by its systematic nature. Everyone has some general conception of the nature of the world in which they live and of their place in it. Metaphysics replaces the non-argued assumptions embodied in such a conception with a rational and organized body of beliefs about the world as a whole. Everyone has occasion to doubt and question beliefs, their own or those of others, with more or less success and without any theory of what they are doing. Epistemology seeks by argument to make explicit the rules of correct belief formation. Everyone governs their conduct by directing it to desired or valued ends. Ethics, or moral philosophy, in its most inclusive sense, seeks to articulate, in rationally systematic form, the rules or principles involved."

In order to understand the significance of philosophy in engineering knowledge emergence, types, renewability and dynamism, it is necessary to save them from steadiness by philosophical thinking; critical criticism and the foundations must be based on rational thinking level. In well-established and generative engineering institutions, engineering education is not based on memorization or transportation of information and knowledge without criticism but the weight is given to more interrogation about their contents and functions with unsteady development. It is, therefore, necessary for an engineer to have such a dynamic thinking capability prior to the intake of ready spoon fed frozen information. Hence, the emergence of healthy and productive thinking and physical functioning are possible in a sustainable manner not for the engineers only but also for other specialists, in particular, and the society, in general.

On the other hand, philosophy is the love of wisdom or knowledge, but wisdom is recognizing right from wrong, while knowledge is recognizing truth from falsity; two very different concerns which belong to two separate institutions, religion and science. Understanding is the combination of wisdom and knowledge, which implies that understanding, is within the domain of philosophy.

3.3 Uncertainty and Fuzziness of Philosophy

One of the powerful uncertainty and fuzziness sources in human ever-emerging desires of various kinds is from simple physical desires, which are shared with other animals to much more complicated desires specific for human nature. Every desire agitates the mind and distracts the process of concentration indispensable for an act of understanding to be productive. The stronger an emergent desire, the higher the degree of agitation it stirs up, the less the degree of concentration of mind; and the less the degree of concentration, the fuzzier the process of thinking, the lower the degree of understanding and reduction in the uncertainty without its complete isolation. Most of the desires self-propel their intensity—the more one tries to satisfy them, the higher become the demand; the way of moderation—the 'middle way' as in the Hadith of Prophet Mohammad (pbuh), is hard to follow when the fire of desires is burning inside us and making the minds restless, turbulent and obstinate. Amidst of such feelings, the human mind is completely free in thinking including every extreme towards any direction with uncertainty ingredients. This is referred to as the philosophical thinking, which must be filtered later through the logic rules for deducing proper, meaningful and useful statements (arguments) leading to plausible conclusions, but even then the uncertainty element cannot be driven out absolutely.

The restlessness and turbulence of minds are permanently intensified by the stress in which one lives due to the competitiveness inherent in today's society and the helplessness of majority of us to get out of the social boxes and cages (in which we have been pushed by economic forces too strong to withstand), even if we desperately desire to.

Although the strength of passion with which one may pursue truth and understanding is a powerful stimulator and 'energizer' of thinking, understanding needs 'peace of mind'—a mind, which is calm and cool, composed and collected. Paradoxically enough, while being sources of fuzziness, mind and desires are, at the same time, key factors for overcoming (transcending) it, especially if it relates to problems deeply rooted in human experience, which differs from individual to others, although there is a gross overlap. Non-overlapping portions are full of uncertainty and fuzziness in linguistically terms [5].

The uncertainty and fuzzy concepts in understanding problems that emerge out of life complexity as it unfolds cannot be resolved at the same level of knowledge that one has when these problems appear. Only when one's consciousness is expanded i.e. raised to a higher level, then the tension fades and the problems, being seen in a new light, are no longer problems. When problems dissolve, one may say that the fuzziness related to them has been transcended, but this is on relativistic sense, not in absolute sense.

3.4 Philosophical Inferences in Engineering

After a set of observations and their approximate reasoning one can reach to conclusive inference in an inductive manner. Irrespective of whether the conclusions are correct or incorrect according to crisp inference (two-valued logic, correct to a certain degree under the light of fuzzy logic) and probabilistically correct in some certain situations. Even though there may be a common conclusion at the time, but it must be verified with additional observations, methodological improvements and modern models.

A Bayesian inference to a problem starts with the formulation of a model with hopes that it is adequate to describe the situation of interest initially. A prior probability distribution function (pdf) is then suggested over the unknown model parameters, which is meant to capture one's beliefs about the situation without data. However, with the incoming of data one may then apply Bayesian inference rule to obtain a posterior distribution for the same unknown parameters, which take into account the prior pdf and the data. From the posterior distribution, one can then compute updated predictive pdf for future observations [6]. The Bayesian approach can be simply applied and justified theoretically as the proper approach to uncertain inference by various arguments involving consistency with clear principles of rationality. Even though, a prior pdf selection seems subjective, but it is not arbitrary. It is necessary that the priory pdf should capture one's correct prior information by taking into consideration a combination of prior beliefs.

In the past, engineering was under the concept of architect, who is concerned with linguistic reasoning and not mathematical equations as engineers today. It may be said that architects are more philosophy oriented than engineers. Architects care for comparatively very less mathematical principles and science because they regard themselves more artistic oriented. In engineering education system mathematics, physics and chemistry are among the basic courses, but these hardly exist in many architectural institutions in the world, except in developing or underdeveloped societies. Also in engineering, there are many calculus works, which are neither in architecture nor in philosophy. For instance architectures are thought free-hand drawings and sketching, but engineers are more regular shape drawing oriented without free-hand works. In the architecture education system although the history of architecture and site or field trips to old structures are significant, in engineering these are ignored completely.

Engineering and technology are applied science and they have some overlapping with philosophy, which should be enlarged in future. The author suggests that without philosophical debates (rational linguistic discussions) how can the ideas expand? And subsequently how the logical inferences can be deduced? Of course, any theory, hypothesis, equation or algorithm has linguistic background garden full of with minor and major ideas, some are beneficial and others are not relevant to the current point of interest. Bo-cong [1] suggested science, technology and engineering as trichotomy, where there is a need for separate philosophical issues in each of these disciplined. Since engineering is in close contact with economic, social, management, partly political, ethical and psychological factors, what type of cement can mix them to reach a beneficial product? The answer is philosophy followed by logical inference. It is not always that technological factors play role in engineering activities, but there are many instances, where non-technological aspects play role also. For instance, we care for educated people as engineers, however, during the history of engineering (unfortunately, history of engineering is also available in a very limited extend) inventors, who are not even graduates of primary school may also come up with some technological idea, which may not be very extensive but may add slight improvement into the existing mechanisms. For instance, reinforced concrete was invented by a farmer, who used to make vases from cement to plant flowers. He noticed that after some time all the vases had cracks and fractures. Then he had the idea of imbedding iron wires inside the cement and noticed that the vases had no more cracks or fractures. This point has

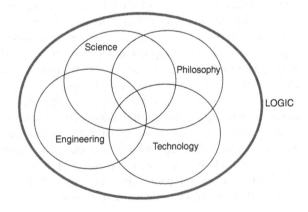

Fig. 3.1 Interactive discourses between various aspects

been discussed with neighbors, and finally, reached the civil engineering chamber and then onwards with further philosophy, rational thinking and logical inferences based on the experiments, reinforced concrete started to be used in civil engineering structures and they are everywhere today.

Although science and technology intermingle with philosophy [7], engineering has rather limited overlap, which must be enhanced in future [8] Fig. 3.1 is a simple schematic correspondence between philosophy, science, technology and engineering, which gain meaning and reasoning under the umbrella of logic. One may question why the logic encapsulated all other disciplines, because conventionally after the philosophical aspects logic is employed for trimming irrational propositions to infer final decision. This is due to the view taken in this book that philosophy here is not a general topic such as inclusion of genuine ontology and metaphysics, but dominantly logical aspects and aesthetics to a limited extend only. Another implication of logic as circulation in this figure is that all the disciplines will contribute jointly or individually towards rational deductions, which are possible through the logical principles.

Today engineering benefits widely from the end productions of scientific works almost without reasoning for direct and ready solutions to various problems, whereas philosophy remains marginal and almost non-existing in engineering creative works. In fact, engineering should be a convenient mixture of different disciplines and the best engineering training should adapt the relationships in different proportions between engineering, science, philosophy and technology. Engineering candidates can have their subjective shares from philosophy, science and technology according their creative capacities. Such subjectivity is unavoidable and very useful for differences in opinion about a problem, so that they can debate, discuss and reach to a common solution at the end with the contribution from different views. On the contrary, especially scientific and technological developments are also dependent on engineering innovations through various instruments and experiments. For instance, magnetic resonance (MR) equipment is developed by engineers for the service of different disciplines among which is the medicine. The most important aspect of engineering is design, which differentiates engineering from science and technology. It is well recognized by many that science is concerned with discovery, technology with invention, whereas engineering is more craft work concerned with making, producing and generating alternative solutions for a given problem. Critical scientific knowledge including theories falls within the domain of science, which is in a continuous development since many centuries. Patterns and blueprints are invention imprints in technological developments. However, engineering tasks are concerned with material products and designs.

Philosophy of science has active role in scientific studies since the science became rather independently spelled out from the philosophy during the renaissance period. During the last few decades, through patent institutions philosophy of technology started to ripen up and there are many articles in the literature about such aspects. However, philosophy of engineering is a very recent debate in the world since the last several years [9]. Philosophy of science remained as an overlooked or delayed aspect in systematic engineering creative thinking for many decades. This does not mean that engineers never benefited from the philosophical principles, of course they did, but this remained on individual basis and it could not be systematized in education curriculums especially during the engineering education (Chap. 6). Ethical behaviors and career rules are based on philosophical basis, but real mass of engineering background on tangible aspects are far away from systematic philosophical principles. Even the entrance of ethics into the engineering domain is a work achieved since few decades only [10]. One can easily say that philosophy of engineering is almost non-existent. Another vague entrance of philosophical thinking into engineering domain may be through technology as in Fig. 3.1. Generally, engineering and technology are thought as distinct disciplines, but there are occasional interferences between the two, which transfer some philosophical aspects into engineering thinking, because some engineers take active involvement in technological developments such as the first atom bomb production. Some may propose that there is no need for separate philosophy principles from engineering, because they are included in the philosophy of technology, but this is not acceptable completely due to the benefit of engineers from the end products of science rather than technology in their mental formation outputs such as design, solutions to many problems and in report writings. Most of the engineering students in the world cannot write proper reports or articles due to their inefficient philosophical backgrounds.

After all what are explained above, one can understand that philosophy of engineering is a virgin topic at the verge of development, and there is not commonly agreed set of fundamentals, principles and rules for the definition of this field. Hence, it is within the context of this book to propose such a set of items, which can be elaborated on more in future.

For the development of engineering philosophy, engineers must be ready to deal with philosophical linguistic explanations of designs, models, formulations, problem solutions, etc. not numerically but initially verbally. On the other hand, philosophers should recognize existence of engineering works, their empirical solutions and decision makings. However, rather than philosophers, engineers themselves should try and provide philosophical grounds in problem solving prior to symbolic logic that leads to complicated equations, the meaning of which may not be clear for engineers who come across with such formulations. It is stressed in this book that if the linguistic background and foundations of any engineering problem are known linguistically (philosophy and logic) then their translation to mathematical expressions is only a matter of conversion from the language (Turkish, English, Chinese, Arabic, French, etc.) to mathematical symbols. Of course, the reverse should also be applicable provided that the meaning of each symbol is explained. Engineer should be able to translate formulations into linguistic compositions, where philosophy (at least science philosophy) and logic prevail.

Apart from classical ethical and aesthetical aspects of philosophical ingredients in engineering, some other issues are also related to engineering, but they are not very obvious or not used frequently. Since philosophy is verbal in character, it is necessary to consider epistemological aspects of each terminology, question the nature of engineering knowledge and its justification in final decision making (Chap. 2). Since epistemology is a sub-branch of philosophy, engineering is related to philosophy through such a branch.

Almost all the engineering methodologies are heuristic in nature, which require accuracy and justification characteristics. Additionally, these methodologies are crisp and based on a set of assumptions and hypothesis, which provide a philosophical basis for further criticism and debate. Since critics, debates, discussions and comments are all in verbal domain; their proper and effective uses necessitate philosophical grounds.

Verbal information and knowledge are capable to produce symbolic expressions that fall within the domain of mathematics. Philosophers should pay attention to the reality of engineering practices based on analytical and especially empirical grounds. Immediate conclusive expectations by engineers do not give way to the entrance of the philosophical issues into engineering aspects. Most often without verbal questioning, interpretation or engineering criticism any engineer seeks ready crisp answers by using one of the methodologies through symbols in equation forms. This is tantamount to saying that crisp thinking and expectations in engineering prevent entrance of philosophical aspects into engineering activities. Any engineer should consider that exact solutions are not possible, and hence, the results must be questioned from different facets. For instance, the "safety factor" concept kills the possibility of philosophical issues to enter into engineering studies and it is rather an "ignorance" factor; ignorance in the sense that all blame is thrown over this factor without further pondering or reasoning in search for a solution in the domain of uncertainty (Chap. 2). One can state that the crisper the engineering works, the less philosophical are the end products in engineering.

Recent movements towards the philosophy of engineering indicate that engineering career started to face in an increasing manner a global crisis that spurs although temporarily a turn to philosophical fundamentals, logical propositions and inferences prior to any quantification.

3.5 Engineering and Design Philosophy

An engineering design may be defined as that socio-economic activity by which scientific, engineering and behavioral principles are applied together with technical information and experience including skill, imagine action and judgment in the creation of functional economical, aesthetically pleasing, and environmentally acceptable devices, processes, or systems for the benefit of society. The followings are among the distinguishing points of the engineering design.

- Recognition of a need;
- Statement of the problem, identification of performance objectives, and design issues;
- Collection of knowledge and information;

- Concept formulation in accordance with the design criteria, search for a method, theory, model, or hypothesis;
- Analysis of solution components;
- Synthesis to create detailed alternative solutions;
- Evaluation of ideas and solutions;
- Optimization;
- Recommendation and communication;
- Implementation.

The philosophy of engineering should concentrate first on the ability of how engineers think in front of problems, how they provide solutions and what is their works' impact on the society, in general. Additionally, rational and logical thinking after a certain philosophical argument, engineers should care about ethics and aesthetics and to a limited extent epistemology as a part of science philosophy. Engineering may modify the environment through the design and subsequent manufactures of artifacts. The philosophy of engineering is the consideration of philosophical issues as they apply to engineering. Such issues might include the objectivity of experiments, the ethics of engineering activity in the workplace and in society, the aesthetics of engineered artifacts, etc.

Throughout the engineering works and studies so far, one can see that philosophy in engineering circles is ignored, and consequently, engineering as a career remained strictly numerical with general linguistically assessments. On the other hand, philosophers did not take into account engineering aspects and it is a common impression that engineers cannot be scientists. Such a conclusion is due to the absence of philosophical and especially scientific philosophical thinking ingredients in engineering curriculums. Philosophy has remained as a hidden and thin layer in engineering understanding and productions. Unfortunately, there is no course related to philosophical thinking principles in any engineering education system and most often engineers are acquainted with formulations, algorithms or flow charts, etc. As a hidden layer, philosophical thinking (linguistically expressions with reasoning) exists in any engineer's mind at different grades and the most beneficial facets of such a thinking can be classified into groups as follows.

1. Philosophy helps engineers to enter philosophical discussions, debates and view exchanges even on daily matters.
2. Engineering career has its ethical rules and regulations, which are all verbal and have philosophical aspects that cannot be expressed in terms of equations.
3. Philosophy, especially philosophy of science, helps engineers to understand the fundamentals of the solutions rather than substitution of crisp data into well documented equations for the solution.
4. Only philosophical thinking may generate alternative solutions for the same problem, and hence, in the decision making situation, an engineer should ponder about each alternative and choose the most convenient one without equations, algorithms, software, etc., but by reasoning based on certain logical rules.

5. Through the philosophical principles engineers can train themselves as self-understanding individuals again with reasoning rather than ready spoon fed information and equation dependence.
6. Philosophy paves way to verbal information domain, and hence, engineer can gain self-confidence in convincing other individuals through discussions.
7. Engineering philosophy gives to engineer ability of conveying his/her thoughts through a common language prior to mathematical equations and crisp knowledge.
8. Present day engineering remains in a rigid domain through sole mathematical and numerical concepts and aspects, which do not help to collaborate with other careers and even with other engineers.
9. Engineering careers have been triggered by scientific philosophical thinking even unconsciously, because in the history many preliminary structures are constructed without equations, currently available theories or mathematical expressions, but reasoning and linguistically rules led engineers towards their final targets with achievements.
10. Life long experience teaches engineers problem solutions that cannot be obtained through crisp mathematical equations. As they become more experienced, they gain expertise, and hence, become experts not due to extensive mathematical derivations, but due to more linguistically thinking and reasoning with experience.

3.6 Engineering and Philosophy of Science

Engineering career evolution has gone through many stages even during extinct civilizations and a specific evolution is taking place today. It is not possible to think of any civilization without engineering works such as bridges, roads and highways, dams, airports, water distribution systems, construction material, earthquake resistant structure design, and weaponry. Initially, engineers had their trainings in the form of master-apprentice information transfer with linguistic (verbal) statements, which had logical, rational and philosophical (uncertainty) aspects all together for the problem solution. Engineers were also in the domain of art rather than science and there was hardly a distinction between an engineer and architect. Today, although architects are still in the linguistic and aesthetical domains, but engineers shifted towards more crisp ideas, numerical methods, procedures, algorithms and ready software for the problem solutions. Hence, rather than creative thinking and specific production, similar case studies and formulated solutions become fashionable, which decelerated the creative and productive aspects of engineering education and training. Old fashion master-apprentice relationship broke and gave way to academic master degree and in the meantime apprentice side either faded away or it is ignored completely.

Prior to detailed information about the importance of the science philosophy, engineering reflections on the sayings of several scientists, philosophers and engineers should be taken into consideration.

> It is the mark of an educated man to look for precision in each class of things just so far as the nature of the subjects admits (Aristotle, 384–323 BC).

> Any knowledge between without application is between the truth and false" (Abou-l Iz Al-Jazari, 1202—Şen [11]).

Today engineers as educated men and women should look for different classifications that are rather subjective, but have common information about other engineers to criticize and try to improve the available knowledge.

> Scientists explore what is; engineers create what has never been [12].

This last statement implies that engineers are not scientists but they should benefit from the exploration of the scientists for creative ideas in improving the comfort of the society. Present day engineering education systems try to empower candidates with physical and scientific findings without much practical use and the philosophical bases. The statement "engineers cannot be scientists", becomes true if s/he is not aware of philosophical thinking but dependent only on the mathematical equations without critical views. For an engineer to become a scientist, s/he should empower himself/herself with principles of philosophy, subsequent logical propositions and inferences. Philosophy and logic necessitate linguistic (verbal) means, which are missing in engineering education systems; instead present day engineering trainings are full of numerical formulations, symbols, equations and software. The first intensive contact of engineers with logical principles is due to software writings, where any slight logical error causes mistakes and the software do not produce desired outputs. Debugging in any software development requires science philosophy and especially logical principles more than any other numerical calculations in engineering domain.

> The origin of the science of classification goes back to the writings of the ancient Greeks but the process of classification, the recognition of similarities and the grouping of objects based thereon, dates back to primitive man [13].

Systematic scientific writings that started with Old Greeks were based on the philosophical thinking and rational bases as a result of knowledge and information accumulations from ancient civilizations, which are ripened more with the rational thinking principles. Original and creative ideas have started with philosophical thinking in deductive and to a lesser extent inductive ways. Ideas are criticized, discussed and rendered into improvements for the service of the society. In ancient times, there was no distinction between a philosopher, scientist, engineer and an architect. Any individual had philosophical thinking, logical deductions and rational bases at different grades, and consequently, according today's criteria s/he can be specified as an engineer, scientist, philosopher or an architect only. As a trend from the engineering history, engineers should have philosophical bases in their creative works, but not pure philosophy, instead science philosophy. One should ask at this stage, which type of philosophy should be given during the engineering education today? Is it pure or science philosophy? Since, engineers are concerned with the comfort of the society; in general, leading to objective solutions, science philosophy is necessary in their basic education program so that they can adjust ideas linguistically in a qualitative way prior to numerical solutions. In this way, engineers will be empowered to suggest not only a single solution similar to case studies or problem solutions in text books, but several alternatives, in which case the engineer then tries to select the most rapid, cheap, secure, and optimum solution by reserving other features for future use.

> The mere formulation of a problem is far more often essential than its solution: to raise new questions, new possibilities, require creative imagination and marks real advances in science (Albert Einstein).

The above saying indicates that none of the formulations provide a unique solution of the problem at hand, but approximate results. This implies that any formulation has improvement possibilities provided that the scientist or engineer wants to think analytically with the support of science philosophy. Engineers after their 4 year Bachelor of Science education, depending on their analytical thinking abilities, try to solve problems according to readily available formulations or software without creative thinking capability. For an engineer, philosophical thinking means to understand foundations of engineering problems not through the symbolic logic and symbols as in the formulations, but their logical rules [14].

> If you can measure what you are speaking about and express it in numbers,
> you know something about it (Lord Kelvin).

This statement suits to quantitative physical and engineering aspects, but it also implies that prior to numbers, the mathematical expressions should be in verbal information forms. Any instrument provides measurements, but without knowing the possible scale domain of the measured variable, it is not possible to accept the measurements straight out as accurate and useable identity in engineering problem solutions.

Recent movements towards the philosophy of engineering indicate that engineering career started to face in an increasing manner a global crisis that spurs although temporarily, a turn to philosophical fundamentals, principles, logical propositions and inferences prior to any quantification.

Almost all the engineering methodologies are heuristic in nature, which require accuracy and justification characteristics. Additionally, these methodologies are crisp and based on a set of unrealistic or approximate assumptions and hypothesis, which provide a philosophical basis for further criticism and debate, which are all in verbal domain; their proper and effective uses necessitate philosophical grounds.

Philosophy triggers desires for generation of innovative inventions and knowledge. However, spiritual thoughts (based on culture, civilization, language, religion, ideology, patriotism, etc.) differ from person to person without common objectivity, but their support to rational thinking may provide additional internal energy and excitement. Many individuals, institutions, associations and establishments may mention differently about critical criticism, analytical thinking, global approximations, science and topics like understanding the nature, but they do not mention about the common base of all these points, which fall within the circle of philosophy. If philosophy is exempted from these components as cement, the research spirit of human might be boring again indulging to repetitive and ready memorization rules. For instance, many may think that for critical debate, internet connection may be useful, because it is information technology for knowledge transfer. Anybody can think that s/he can draw useful information from internet and may print them on paper keeping in mind that s/he will read it later in detail. In this manner piles of paper lay next to him/her. Is it possible to deal with them without philosophy and critical assessments? Or is it better to intake knowledge simultaneously with philosophical and logical foundations by giving rise to their accumulation in minds rather than on papers? It is a personal experience that after some time, the accumulation of papers is useful only to use their reverse sides as scraps. Internet information cannot be safe completely, and therefore, it is necessary to verify their validity by philosophical, rational and logical means. In order to keep critical thinking on line and alive, one should make critical revisions frequently. Philosophy teaches the principles of critical thought. An engineer empowered with such critical view abilities can revise any information with rational, philosophical and logical rules. S/he can then render the intake of information into useful forms for the society. This information remains in his/her memory linguistically for any future verbal, symbolic and formulation generations. Convenient forms may be given to logically base engineering philosophical knowledge in any problem solution. If the thought is critical then reading, writing and productions also become critical and alive. One cannot say that there remains no memorization, but engineers equipped with philosophy and logic can reserve this in their mind even after their repetitive usages with dynamism and triggering at times of need. Another property of an engineer is a set of practical thoughts and solutions and these are also provided by philosophical approaches. After raising philosophical engineering knowledge to the level of practicality, it can be employed by engineers in convenient cases for solutions. Such practical knowledge can be transferred to new engineering adults as simple as possible with principles of philosophy of engineering. This gives to engineers the ability of feeling that s/he can also generate information and use knowledge at proper times and locations for problem solutions.

In classical engineering education or even during the secondary school training, there may be some memorized information, which may be rendered into more active and the least static status by philosophy. The most significant separation of the philosophy from engineering is that, in philosophy, there are no ends for solutions, whereas in engineering there must be ends for applications. Any mind acquaintant with philosophy tries to find not a unique solution to the problem at hand, but several alternative solutions. If engineering is defined as the practical application of available scientific knowledge, then such a definition may lead to undesirable conclusions as if engineers do not need philosophy. This consequent may be documented on the basis of scientific philosophy, because science has more rigid rules compared to philosophy and philosophy of engineering is akin to the philosophy of science.

After all what have been explained in the previous paragraphs, it is the main purpose of this book to reflect scientific philosophical aspects that are necessary in the engineering education (Chap. 6). In the past, master-apprentice training for an expert engineer has become more involved in the universities as if apprentice stage corresponds to "Bachelor of Science", master stage to "Master of Science", and finally, expert level can be viewed as the "Philosophy of Doctorate". In these three stages of modern education system, "science" and "philosophy" are emphasized even in engineering training. The graduates seem to have been empowered with analytical thinking capability, which helps engineers to memorize, transfer and ready use of knowledge according to past applications. It is emphasized in this book that science and its philosophical foundations should be given to engineers for better problem solving and even personal emotional and intuitive comfort help to improve practical and creative intellects.

3.7 Basic Engineering

Engineering and architectural thoughts are related to metaphysical and ontological issues such as imagination, description and visualization of a certain problem along the solution path, and therefore, philosophical and logical interpretations and inferences are necessary to reach at a final solution among many alternatives after an effective decision making process. No need to say that such philosophical ingredients in engineering are relatively rare compared to the methodology, epistemology, ethics and aesthetics. A common ground among the engineers and philosophers include topics of ethics, aesthetics, epistemology, methodology and ontology. However, these two groups of specializations have shown common interest during few decades only on ethics and since many centuries on aesthetics, but not so on other issues. Perhaps, the first impression in both specializations is towards looking onto each other as if the interests are different, but as long as verbal issues are concerned, engineers must shift towards the philosophical domain so as to increase their ability to draw deeper conclusions for their current problems prior to numerical solutions. Coupled with the preassembled linguistic (philosophical and logical) understanding, any formulation or equation appears as a matter of dynamic activity on the engineer's behalf.

Engineering can be defined also as an art and ability of rendering available natural resources for the service of human after scientific inferences and their end applications. In this statement, there are two words, "ability" and "art", which can be explained only philosophically. Scientific theories and results are produced by scientists and as end product users, engineers by acquainting themselves through philosophy of science principles may become along with scientists closely with common and joint agreements. After all, scientists who are empowered with the philosophy of science and any engineer with background on the philosophy of engineering can find a common hinterland for discussion. Unfortunately, without philosophy of engineering, engineers are confined to blind and restrictive applications of scientific outputs only. In order to avoid such a situation, engineers must be empowered with philosophical thoughts also in addition to the ethical and aesthetical implementations.

Classical engineering solutions do not consider the philosophical principles but recently mostly economy. Such solutions should not be considered within wholeness; otherwise they may have harmful ingredients. Instead of trying to improve the occurrence of harmful cases afterwards, it is far better to search for solutions by considering philosophy of engineering principles from different facets and suggest an engineering solution based on linguistic information with subsequent logical rules. Mathematical formulations have logical linguistic implications at their bases, and hence, the engineer must put forward their thoughts into action linguistically with a set of knowledge about the subject concerned. Philosophical knowledge provides reasonable distinction between the useful and harmful aspects of a problem, and hence, useful aspects may be selected rationally. Today engineering curriculums include socio-economic and cultural courses for linguistic training of engineers, but without philosophy of engineering such a desire may not reach its target successfully. In engineering all the symbolic and numerical results can be useful only if they find their linguistic fundamentals in engineers mind and memory. Additionally, philosophy of engineering principles is bound to provide a more dynamic basis for better understanding of socio-economic and cultural courses. Luegenbiehl [15] defines engineering as:

> The transformation of the natural world, using scientific principles and mathematics, in order to achieve some desired practical end.

In this definition, the words "scientific" and "mathematics" as preliminary requirements for the existence of engineering implies "philosophy of science" and "logic", which are disciplines under the philosophy. Unfortunately, many engineering institutions all over the world do not care for philosophy of science or logic. They drive away these two major legs of the modern engineering career and concentrate more on science for the sake of science and mathematics for the sake of knowing mathematics. Consequently, engineers seek solutions to their problems by using the end products of the scientific achievements and mathematical end products in terms of formulations, equations or algorithms without linguistic basis, where philosophy can provide a creative thinking domain. They know superficially that mathematics means logical principles and science finds its basis in the philosophy. Mitcham [16] gives the general characteristics of engineering through a more linguistically side as follows.

> What engineering is might be better determined by how the word "engineering" and its cognates and associated terms (such as invention, innovation, design, technology, science, etc.) are used, especially in relation to each other. From a linguistic philosophical perspective, it would be appropriate to begin not so much with our experiences of engineering, but with the words we use to talk about such experiences.

On the other hand, Davis [10] is against the idea of philosophical definitions of engineering and a linguistic approach. He suggests engineering definition as,

> All attempts at philosophical definition will: (a) be circular (that is, use "engineering" or a synonym or equally troublesome term); (b) be open to serious counter-examples

(whatever because they exclude from engineering activities clearly belonging or because they include activities clearly not belonging; (c) be too abstract to be informative; or (d) suffer a combination of these errors.

After such criticisms he suggests engineering definition from the historical point of view as follows.

Engineering, like other professions, is self-defining (in something other than the classical sense of definition). There is a core, more or less fixed by history at any given time, which determines what is engineering and what is not. This historical core, a set of living practitioners who—by discipline, occupation, and profession—undoubtedly are engineers, constitute the professions.

3.8 Philosophy of Engineering

In the past, bridges, aqueducts, mosques, cathedrals have been built; say before 500 years without numerical solutions. However, is it possible that present day engineering structures can survive many centuries without philosophy of engineering aesthetics, durability, and economic consequences in the long-run? Fantastic examples of such structures can be found today in many places and also in Istanbul, Turkey, as St. Sophia Museum from the fourth century after Christ, and Blue Most from the fifteenth century as well as Roman aqueducts in the center of Istanbul in addition to city walls from Byzantine period. In 1998 there was a severe flood that occurred on the European side of Istanbul in one of the valleys. After the flood occurrence in Istanbul, Turkey, one could realize that bridges by Mimar (Architect) Sinan (1490–1588) from sixteenth century were intact, whereas recently made engineering bridges were all demolished or played the role of small dams with backwater effects, and therefore, surrounding areas were subjected to inundation. Ancient engineers and architects did not have scientific tools such as theories, equations, formulations, crisp algorithms, computers and software, but they depended on their philosophical thoughts, imaginations, verbal information, and logical rule bases, and consequently, they based their designs on these knowledge sets leading to rational solutions. However, present day engineers, without resorting to engineering philosophy concepts, found numerical results from well-established equations (are they really valid?) or ready software and applied them perhaps without linguistic (philosophical) judgments and critical reasoning.

Today one may witness many environmental problems, which may be there, due to the absence of the philosophical (ethical, aesthetical, logical, etc.) principles and ready applications of crisp formulations without interpretations and possible consequent assessments of the aftermath phenomenon. One must not forget that any engineering formulation is with assumptions, simplifications and idealizations, all of which may not be suitable for the present problem at hand. Hence, any engineer with philosophy of engineering and logical principles can suggest alternative formulations to present cases for solution. Not static knowledge only, but dynamic philosophy of engineering principles makes an engineer alive, active and successful in his/her career.

Philosophers may not provide a separate position for engineers in thought system. However, engineers must try and benefit from philosophical fundamentals and principles in their work productions under the physical rules and try to direct their thought productions towards betterment of engineering activity. In this way, engineers may reach to more useful products as arts and admirable structures. For instance, many years ago I had my education in engineering without any philosophical principles, let along philosophy of engineering, which is a recent emergence. I could judge logically and verbally (philosophically) all the formulations and equations that were accumulated during engineering education after learning logical and philosophical principles many years later, which lead to more criticism, skepticism, and rather than verification falsification of the findings become effective as Popper [17] and Reichenbach [18] insisted. Today in engineering education curriculum social content courses are given but still logic and philosophy related subjects do not exist. Without philosophy of engineering and logical rule base generation all of the formulations in the memory of an engineer will remain there statically without any dynamism, and hence, engineers will have rather stagnant, non-generative and non-creative mind. Any engineering education without philosophy of science and/or technology, let along philosophy of engineers, is bound to provide textbook example solutions without any creativity. If engineers or perhaps philosophers gave significance to philosophy in engineering training, for instance, even at the graduate level, then engineers will be acquainted with more creative thinking capabilities, because they could then criticize even the knowledge that they take from the instructors. However, philosophers cannot be accused for this, because the philosophical fundamentals and principles are available for engineers to use them. Philosophy is an integrated (whole) thought system whereas science, technology, art and engineering are under the umbrella of philosophy and each one has its share accordingly. Why then engineers did not benefit from such shares? It is possible to answer this question according to each culture or society. However, in many countries there are not philosophical excerpts in the curriculum. Perhaps, engineering education system is a discipline, which has not benefited from philosophical facts at all. All over the world, anybody with Dr. title has gone through a systematic training ending with a thesis, which is entitled as "Philosophy of Doctorate", where "philosophy" is spelled out explicitly, but the holder is not trained or empowered with the philosophical ingredients. This title ensures that the holder should be able to explain his/her specialty topic philosophically (verbally, linguistically), but generally the text is written according to a language without or very little philosophical criticisms.

Although scientists and engineers alike try to improve human security and comfort, but in the meantime the methodologies, techniques or instruments can be used destructively. Herein, philosophical sub-divisions as ethics and aesthetics come into view as interrogatives. Even noise pollution from any engineering production can be considered as an ethical inconvenience for the society. If philosophical principles are not cared for in engineering, then engineers may try to save the day only through their stagnant and traditional knowledge embedded in their minds. Hence, they may not heed for criticism, questioning, interrogation, and hence, they may apply their memorized prescriptions without benefit. Most often a formulation, procedure or software is not criticized by engineers prior to their applications as for their suitability for the current problem. If the fundamentals of knowledge are not taken on philosophical and logical foundations, then the engineer may not even know how to achieve the work dynamically.

Since, philosophy is a way of rational explanation of natural objects and any phenomena, an engineer apart from the understanding of a scientist, must draw a share from this definition. Even though engineers are not scientists, they may come close to scientific findings by introducing themselves with the philosophy of engineering. It is possible that an engineer can reach depths of scientific knowledge provided that s/he is interested in interrogation, criticism and interpretation of the end products from different angles on philosophical and logical bases. S/he should do all these for a successful application of the knowledge in a dynamic manner. I had an engineering background, but after graduation many years later, I understood that engineering education in its current status without philosophy of engineering and logical rules cannot give engineer ability to come closer to a scientist. In various discussions with many scientists, who did not have proper engineering background, in different countries, I entered into scientific discussions about what is the science? What are its principles? What are the features of scientists? We came to the point that "engineers are not scientists". When others realized that I have engineering background, then they mentioned about exceptions. Such an exceptional statue is gained by trying to acquaint oneself with philosophy of science and logical rule generation verbally rather than numerically. Such extra interests give one ambition to work on much interestingly even after graduation and at the end, one also may came to the conclusion that "engineers cannot be scientists" until unless they have acquaintance with philosophy, in general, and philosophy of science and philosophy of engineering in particular. Philosophy of science covers many aspects. Linguistic information, on the bases of philosophy and logic, can provide mathematical and engineering formulations or equations, but the reverse is not true and leads to a dead end. Philosophy is not the property of any career; it has inter-career and service characteristics without distinction. Unfortunately, in the classical education systems, philosophy of engineering principles are overlooked and engineering is defined only on the bases of economy, simplicity, speediness, practicality, etc. Crisp engineering solutions may overlook environmental conditions and at the end harmful productions can take place without improvement after destruction. Presently, greenhouse effect, global warming and climate change are among such phenomena.

Today we witness many environmental problems, which may be there due to non-existence of the philosophical (ethical, aesthetical, logical, etc.) principles but ready application of crisp formulations without interpretations and possible consequence assessments of the aftermath phenomenon. One must not forget that any engineering formulation is not without assumptions, simplifications and idealizations, all of which may not be suitable for the solution of the present problem at hand. Any engineer with philosophy of engineering thoughts and logical principles can adapt the existing formulation to present situation. Hence, not static knowledge but dynamic philosophy of engineering principles make an engineer alive, active and productive.

There is not a mathematical management system in any education institution but linguistic schemes and flow diagrams exist with explanations for different levels. This may be regarded as the management system (formulation) of the institution concerned. In such a formulation, social responsibility, work power, share, etc. terms and concepts may exist classically and may be expressed by those who work there. However, examination and testing of these concepts and terms can be done either internally by the workers or externally by clients. Such tests may be classical but their philosophical and logical assessments may yield better ideas that may fill the gaps with better improvements. Especially in engineering for joint productive works, philosophy provides a common basis for knowledge generation, ethics and aesthetics.

It is possible to understand that engineering problems of today cannot be controlled with static logic rules (white–black, yes–no, two-values crisp logic or symbolic logic) instead fuzzy logic (i.e. linguistic logic) has been suggested (Chap. 4). In philosophic and logical approaches to various problems, there are not equations or ready formulations but linguistic prepositions and logical inferences. The fundamentals of this approach are human thought, philosophy, logic, interrogation, suspicion, and accordingly verbal rule derivations. In each career, next to the terminology, concepts and prepositions exist according to natural behavior of the phenomenon. The set of such prepositions provide preliminary solutions even approximately. In this way, an engineer puts forward productivity by his/her intellect and linguistic suggestions without static formulations and this gives him/her encouragement and self-reliance. This also helps to generate a synergy without cost and hence one may see that there are relaxations and peace in mind, heart and life leading to knowledge happiness.

Is it possible to have a productive design without philosophy? The answer is definitely no. It may seem at first instance that in engineering education social, economic, cultural and historical dimensions may provide such a situation, but after some test it becomes clear that without philosophy any of such non-numerical courses can achieve its goal. It is obvious that not sayings but core knowledge may have active role by philosophy only, otherwise such knowledge will remain as static in mind. One may think that the culture motivates, but history indicates that culture without philosophy cannot lead to productivity, however, remains local and comparatively are meager.

In engineering there is a misunderstanding as if the first thought stage is mathematical thinking. The main reason for this is that in engineering there are many formulations based on mathematics, but it must not be forgotten that the bases of mathematics are philosophy and logic. Through the mathematical thinking philosophy could enter into the engineering domain. It is of utmost importance to introduce in the engineering education that even the complicated mathematical expressions have rather simple philosophical and logical verbal ingredients (Chap. 6). Is it then sufficient to use scientific findings that are based on philosophy of science in engineering rather than teaching philosophy of engineering principles to engineers? Or is it preferable to teach philosophy and logic principles prior to mathematical formulations? This provides a general linguistic arena for the solution of the problem in terms of logical rules, which can then be translated into mathematical symbols. Many prefer verbal information, which can fire many other verbal thoughts in mind, and hence, leads to productive mental functions. Mathematical thinking becomes useful and productive after the teachings of philosophy principles and logical rules for inference. If engineers want to take their right positions in the development of civilization then they must be trained not only on crisp engineering topics but additionally on social and more significantly on science, technology, engineering philosophy and logic topics. They must not depend only on their career knowledge, but also on the clients' requests by linguistic communication. Only philosophy of engineering can combine effectively both client and engineer opinions collectively towards a common sustainable and satisfactory target.

Engineering communities play far most significant role and have impacts on society more than technological or scientific societies. Philosophers can gain more influence in the society if that pay more attention to engineering issues. Contrarily, engineers should also heed for philosophical issues related to engineering aspects, because engineering activities encompass almost whole activity in a society. Engineers are after practical and efficient end products for the benefit of society whereas philosophers are in search for truth and ontology. Philosophy of engineering should be closely related to sociology of engineering issues. Today, there are sayings as "community engineering", which does not have any deterministic methodology but linguistic debates in fuzzy manner, but it has some end products that are for the benefit of the society concerned. Modern philosopher Popper [17] though that the third world has the essential products of the human mind. This is the reason why many minds travel to centers of scientific, technological and engineering centers especially those in developed countries. Philosopher's reality can be shifted towards engineering not truth but realism by trying to answer to questions of why and how engineering reality comes into existence and the end products are used by society. Philosophy helps engineer to feel what is engineering and why are the engineering products impacts social, economic, physiological and even to a certain extent emotional feelings. Engineering philosophy is necessary because it is not only mechanical ordering of the things, but it is also involved in sociology, economy, politics, management and ethics. Each one of these aspects requires attention of philosophers and sociologists. Humans are discriminated from other creatures as rational animals whereas some gives definition as toolmaking animals (Benjamin Franklin). The former definition is more towards the center of philosophy and the latter is more engineering oriented but still needs philosophical rational thinking. Engineers try to match the harmonious relationship between nature and human. On the other hand they play significant role between the society and individuals. Is it possible to arrange automatically such relationships through mathematical or deterministic types of solutions? Of course, the answer is negative, and therefore, the solution mechanism needs verbal debates and discussions and consequently philosophical thinking enter the scene even though the discussers are engineers. Wisdom is a central issue in the philosophical issues and it tries to lead to a better mode of life not materialistically but also spiritually. It is long known that spiritual feelings in inspirations play important role in rational thinking, and therefore, again engineering has to leave the door partially open for philosophical aspects. Goldmann [19] stated that "Philosophy of engineering should be the paradigm for philosophy of science, rather than reverse".

The subjects of philosophy have broadly three categories as epistemology, metaphysical and ethics, where at least in this last category any career and likewise engineering shares some parts. Philosophy provides engineers to have more reflections on current problems with alternative solutions, which need to linguistic rational distinctions among them for preliminary elimination then numerical procedures may be used for further eliminations until a final decision is made on a single alternative.

There is analytical distinction between science and engineering, because there are special forms of logic in engineering as will be explained later in this book. A proper philosophical analysis of engineering will widen the engineering horizon by rethinking and questioning about the possibility of rethinking based on human common thinking in engineering aspects for the betterment of the final products. Mitcham and Mackey [20] stated that philosophy of engineering will help to criticize and attack misconceptions and false beliefs such as those associated with general theories as those of technological determinism or technological fix—often by arguing their incoherence or dependency on category mistakes.

For instance, computer software although depend on logical principles but prior to this philosophical aspects must also be grasped satisfactorily. Otherwise the programmer cannot lay down the statements in software in a logical and rational manner. This is one of the main reasons why many engineers cannot prepare even simple software, because they are acquainted with solid formulations and equations without philosophical (linguistic knowledge) and logical principles. So how can one expect from an engineer to write logical and rational software without the philosophical fundamentals and logical steps in the generation mechanism of the phenomenon concerned with its input and output components. Furthermore, the basis of analytical philosophy is linguistic philosophy and Wittgenstein suggests that clarification of language use can enabled one to see through certain conundrums that have accumulated in both popular and professional philosophical thought. This linguistic alternative of philosophy helps engineering to open and widen its way for further refined solutions at low cost and fast speed. Unfortunately, epistemological philosophy has paid, if not ignored, little attention towards engineering aspects. This may be due to the case of engineers who does not engender epistemological or logical puzzle like problems as in science domain. In engineering conscious elaborations can ripen only through philosophical thinking and rational logical inferences. One can say that the problems of philosophy are problems of language, and like any other discipline engineering is also dependent on language so it also needs philosophical ingredients. The meaning of knowledge (epistemology) reality (metaphysics) or goodness (ethics) are all within the domain of philosophy and one can appreciate their information content through the philosophical principles. Hence, philosophy helps to resolve disputes among the partners of the same issue provided that they use the same language, because all the debates and final decisions are in the form of language statements, but in engineering after the language stage they are rendered into symbolic or algorithmic forms through systematization of relevant heap of knowledge. The philosophy of engineering should start right from the beginning with the meaning of the word "engineering" and its connotative and associations in terms of invention, design, innovation, technology, science, etc. The philosophy of engineering might be viewed in two categories; at early stages the meanings and information content of each word and terminology without any experience and the second category should focus on the engineering experiences, which vary from one to another individual. The method of description and vague design at preliminary stages of any study should include different sorts of uncertainties, and hence, the philosophical domain becomes available for further debates and discussions. In any engineering work the efficiency and effectiveness play important role. The most serious appeal for philosophy of engineering is the linguistic explanations of events, phenomenon or any other issue, where there are vagueness, incompleteness, suspect and uncertainty so that these adjectives leave room for philosophical debates in engineering. However, philosophy of engineering is in its infancy with promise in the future, but it needs a good care not only by philosophers but more dominantly by engineers. Philosophical ingredient entrances will transform engineering to a new phase after the linguistic foundation and logical inference. Many non-engineers criticize engineering as more or less dogmatic in the sense that it is dependent on deterministic, standard and normalized practices without innovative directions. Although the final decision include singulars, but prior to this stage in engineering there is pluralism and during this phase engineers need to base their discussions on the philosophy of engineering so that the pluralism can be rendered into singulars according to engineering criteria. It is stated by Mitcham and Mackey [20] that the most robust engagement of philosophy in engineering will entail engagement from more than one philosophical perspective. Philosophers themselves represent different schools as much as they represent philosophy. The rich possibilities for philosophy of engineering can be realized only through different philosophies of engineering. The philosophy of engineering, in so far as it is able to highlight problems within the customary ways of thinking both within and about engineering may introduce into engineering a kind of liberating skepticism and wonder regarding engineering that could be especially beneficial to a world such as ours which is increasingly dependent on engineering. This in turn might even make engineering more attractive to some who have shied away from understanding it while mollifying to some degree others who have criticized it Mitcham and Mackey [20].

Consideration of diverse subjects in engineering indicates that it is full of philosophical issues, but unfortunately philosophical principles are buried deeply in this discipline. Among the philosophical tendency subjects of engineering are social, political, ethical, anthropological, metaphysical and ontological questions. Creative design procedures are used by engineers through technological uses and methods of knowledge, which can be enriched by the philosophy of engineering.

In general, conceptual, methodological and epistemological aspects occupy the most of the philosophy related features of engineering. Unfortunately, engineers are confined in almost certainty world even though there are many natural and social events that are within uncertainty world, they do not care for developing their philosophy in design, concepts, terminology, knowledge generation and ethics. Many think that technological knowledge is born from the combination of technology and science, but engineering aspects play very important role in such developments. However, it is not well-known why technology and science philosophies are in the scene without philosophy of engineering. One may suggest that philosophy of engineering can be embedded within the technology and science philosophies, but recent trends and developments indicate the need for philosophy of engineering and its subsequent tail of logical rule bases and final inferences that may be transformed into symbolic logic through equations and formulations. Especially effectiveness and efficiency of engineering thoughts and final productions may attract philosophy of engineering and hence linguistic derivations from the world of uncertainty. Vermaas [21] suggested that precisely the engineering criteria of effectiveness and efficiency provide room for further straightforward analyses of technical functions only and prevent mutual profitable collaboration: philosophical conceptual sophistication becomes for engineers quite quickly unproductive hair-splitting, and engineering pragmatism may become for philosophy conceptual shallowness. Conceptual researches, methodological and epistemological aspects in engineering can best be developed by making it relevant to existing research in philosophy of technology. This statement indicates that there must be a definite relationship between the philosophy of technology and the philosophy of engineering. The concept of technical function was taken as playing a central role in connecting the structural and intentional natures of technical arte-facts, and together with the program's plea to an empirical turn to conduct philosophy with a close focus on engineering practices [22].

Philosophy of knowledge is a wide branch of philosophical issues that addresses engineering also. The understanding procedure of human mind is attached with the knowledge and its critical assessment in epistemology. Intuition of different subjects as knowledge is also a part of phycology and epistemology. Any physical object in the world provides imagination and then knowledge in the minds about there are also non-physical creative imaginations and designs which have descriptive features and they are subjective opening doors for philosophical debates for acceptance by a set of individuals for the benefit of society. Knowing something is almost equivalent with believing in it and when belief enters the knowledge area then philosophical critics appear in its development towards a better direction. It is necessary to have some reasoning in order to distinguish knowledge from sole beliefs that do not provide any benefit to society at large. It is necessary to be able to exhibit a proof or to point to the state of affairs that makes the known proposition true, or it may be merely that one has access to expertise that will confirm the proposition as knowledge. This sentence is subject to many philosophical debates. Engineers usually desire to achieve some objective design and such a desire is frequently incomplete from the point of another individual, and hence, there arises a room for further debate and discussion all of which indicates philosophical ingredients.

Knowledge undermines skepticism, which is also at the foundation of philosophical thinking. The design or project that one thinks about may be very skeptical to others and in order to settle down the scale of skepticism there is a need for extensive criticism through debates and discussions, which all fall under the umbrella of philosophy. Finally, true knowledge is accepted by all parties concerned as the beneficial knowledge. Knowledge is still conceptualized largely as justified true belief. One cannot assist that knowledge is justified true belief, but the assumption underlying most epistemological work is still that knowledge is nearly but not quite justified true belief. This statement leaves vague, incomplete, uncertain and skeptical ingredients in the knowledge content, and hence, fuzzy logic can be used naturally instead of crisp knowledge, which files out these uncertainties through a set of assumptions that are full of engineering activity. However, in philosophical thinking the amount of assumptions is at the minimum or without assumption, and hence, philosophical debates and discussions try to promote knowledge towards perfection, even though it cannot reach to the level of completeness. Epistemology is a foundational subject, which attempts to find the bases upon which one can build an edifice scientific or engineering knowledge. Skeptical problems are within the domain of interest by philosophers and provided that the assumptions are avoided from engineering methodologies they all become skeptical, and therefore, fall within the area of philosophy, but not classical or general philosophy. They must be debated and discussed in the philosophy of engineering branch, which is bound to develop in future along various directions of engineering career.

3.9 Philosophical Thinking Steps

Philosophical thinking is one of the primary abilities for a human being to share integrated life sustenance. Human cannot exist without thinking. S/he grasps not only objects but also supernatural (metaphysical and ontological) knowledge. Grasp with critical reasoning rules provides a basis for production towards the benefit of the society. Not only existence of human being is enough, but additionally and significantly his/her understanding, distinguishing and explanation abilities pave way to further knowledge and information generation in a sustainable manner. In any thinking process one can mention about three stages as imagination, design and knowledge production.

3.9.1 Imagination

This ability defines any human as a creature who may imagine in his/her mind various alternative proposals. It is possible to say that "I imagine, and therefore, I exist" is an ontological thought, which drives individuals to have generative and creative ideas. On the other hand, grasping is important and one can also say that "since I grasp, I exist". Anybody can be dubious (suspicious) about the objects that s/he grasps. This leads to the conclusion that thoughts without imagination cannot exist. Imagination means either intangible existence of some phenomenon by itself or reflections from the real worldly objects into the mind of the human being. Imaginative objects are not necessarily real, but for them to enter the scientific world their obedience to some criteria and measurements must not be overlooked. If one says that in engineering there is nothing imaginary, then s/he is mistaken. Extraordinary and metaphysical things and thoughts are all imaginary beings. They become real provided that the medium and conditions are convenient and suitable. For instance, imaginative thoughts of an engineer to design and construct his/her structure in a better shape, more aesthetic manner, stronger, simpler and cheaper are among the continuous improvement possibilities of final production. An engineer without such imaginations remains with stagnant and rather dogmatic knowledge without knowledge generation ability, and as the time passes away after graduation, s/he might be frustrated in the life. Imagination provides continuous refreshment of mind and such refreshments give rise to fruitful critical rational thoughts by time. According to the present day grasps, engineers are blind to imagination abilities as if there is not such an internal activation and this leads to non-generative ideas, and hence, many engineers resort to memorizing the knowledge without reasoning.

3.9.2 Description

According to an old saying, description is another dimension in the thinking procedure after imagination. Although imaginative phenomena are in virtual media, they should be put into a shape form for descriptions. Herein, description implies geometry of the thing that is under imagination. Human mind is capable to elaborate on the geometry of imaginative objects, which provide for him/her a background for criticism and improvement. Description also includes design and planning. Hence, one may state that after imaginative operations in the mind, geometry plays the most significant part for its more tangible formation. This point shows that prior to any mathematical basis, geometry is the most essential ingredient for the human thought evolution, because human can visualize what s/he thinks in an abstract and conceptual manner that helps them to further ornament the imaginative phenomena. Engineering can be defined not only as classical designer, constructor and decision maker based on common geometrical shapes, but s/he should have an artistic structure and creative ability based on imaginations and description bases even though they may remain in the virtual world. The plans and projects of any engineer are reflections from the virtual visualization of geometrical shapes onto tangible media such as papers or computer screens. Similar to an artist, an engineer can visualize a non-existing work through thoughts in a virtual medium and then onwards produces a real work. In such a procedure, ability and sense gain significance. On the other hand, an engineer benefits from the scientific principles and puts his/her thoughts into practical applications. In addition to the applications as molds, any type of learned knowledge, and especially, equations and formulations should be dubious and they should be located in the mind after critical interrogation.

If instead of the stagnant descriptions when even slightly innovative forms are adopted then engineers might have ambitions towards further stimulations for additional dynamic innovations and inventions. Such an excitation might come from the internal feelings of an individual as a spark or triggering and likewise its appreciation by others may render it into a flame enhancing self-reliance. Any suggestive simple and innovative criticism by others may provide additional thinking dimension for the same individual. Hence, spark and aftermath as thinking production may take the shape of a volcano. Similar to old volcanoes' causative change on the earth surface shape, these thought volcanoes provide ignition of new volcanoes in engineering thinking. Even though volcanoes are explained for engineers in this book, they are equally valid also for other careers.

3.9.3 Production

The productions of an engineer are not only tangible works and structures, but in the same time, they should also be molded into idea productions for the benefit of human comfort, security and health. Since ideas and knowledge are produced in the mind, their transfer to other individuals is possible linguistically after reasonable inferences following interpretations and recommendations. On this respect, it is a wise behavior to benefit from the fundamentals and principles of philosophy, and hence, to intermix engineering with philosophy of engineering. This stage is named as pondering, which leads to useful information and knowledge generation after a steady thinking procedure.

Engineers must perform their productivity by simple and fast ideas in addition to economical solutions not only in the materialistic sense but also in the thinking domain. Productive knowledge and inventions can be shared by other individuals through linguistic expressions rather than equations or formulations, and hence, language is the initial means for such explanations.

3.10 Knowledge Philosophy

Necessary and enough conditions for understanding of any object pass through knowing process. Although perception is a first condition for knowledge, it is not sufficient. Completion of knowledge intake is possible after perception provided that the object is thought upon and its meaning is understood for proper explanations. Accordingly, any person with knowledge does not mean that s/he is knowledgeable. For knowing only, it is sufficient to pass after perception to the stage of mind activity and then to the storage of the object properties rationally in the memory statically. For knowledge, it is necessary to make some activation in the mind after perception. For such activation, interrogation and critical reasoning are necessary in order to end the static stage. If knowledge is stored in the memory without any criticism and rational skepticism, then it renders engineers as knowing individuals but without knowledgeable specialists. Graduates may be loaded with knowledge, but without dynamic use of them. They can use the static knowledge only for classical applications without much mind functions for complete and generative alternative solutions.

In the field of philosophy, the suspect of knowledge, its existence and verification, accuracy degree, etc., must be criticized through debates, comments and opposite views for a better knowledge theory that is known as epistemology. Perceptions are not enough for philosophical thinking, but they are necessary to give meanings, interpretations, explanations, dimensions, etc. Systematically intaken information with criticism provides dynamism in the mind of each individual with an ambition to a certain extent for generation of new ideas and thoughts towards improving human society security and comfort.

3.11 Engineering-Philosophy

Many people conceive that engineering and philosophy do not have common issues and they are crisply separate as two islands in an ocean as shown in Fig. 3.2 Engineers regard themselves as powered with engineering principles mostly mathematics, geometry, physics, mechanics as sciences without attaching any significance to philosophical principles. The engineering concepts are as white and clear as two times two equal to four, whereas philosophical basis is almost black with some general ideas without possible self-defense on these issues.

Engineering is customarily divided into a number of different branches such that each one of these domain has many subdivisions, and especially, on the engineering side the boundaries between the subdivisions are almost impermeable and each specialization thinks that it is a confined compartment that serves to its members only (Chap. 2, Fig. 2.3). Similar pattern is also valid on the philosophy side, but with the boundaries that are not as rigid and impermeable as engineering. However, even after such subdivisions the knowledge exchange is not possible let along between the two islands but also between any two subdivisions. Presently, ethics and aesthetics on the philosophy side are considered as important

Fig. 3.2 Engineering and philosophy in our days

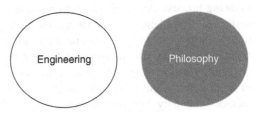

components that influence engineering careers. For instance, safety, risk, and environmental protection questions and their quantification through engineering methodologies are not sufficient in the applications but ethical judgment in assessing their proper influence on design decisions requires philosophy (especially ethics) as an internal practical need of engineering.

Excessive engineering activities have caused several environmental problems such as air, water, atmospheric and various pollutions that are the main nuisances to human race, and consequently, many sectors started to protest engineering activities from ethical, aesthetical, political, and metaphysical, etc., points, which are linguistic and verbal forms without any engineering gadgets such as equations, numeric solutions, algorithms or software. For instance, recent common nuisance against the future development of human is the greenhouse effects; global warming and climate change, which all lead to the deterioration of living standards for all creatures. Many protests against the polluters (mostly engineers) are raised by common or learned people who are most of the time outside the engineering domain, and consciously, or politically, they involve themselves with issues that are damageful to common interest in a blunt philosophical manner. Feminist critics have even associated engineering with patriarchal domination, the death of nature, and the loss of world-centering care [23].

In attempting to define what is meant by a "philosophy" of science, the first problem one encounters is the notorious vagueness of the term "philosophy". A direct consequent of this statement is to raise a question as to how the science itself is objective but its foundation as philosophy is vague, imprecise, blurred and rather uncertain. How can scientific development become possible if the science and its philosophy are uncertain? Most often common man expects or thinks that the science moves toward a unified account of the world but the pictures of reality become ever more disparate. Especially, many scientific theories which were believed to be true turned out to be false or semi-false or there are a lot of debates about their verifications or falsifications. Hence, in the domain of scientific philosophy the scientists become rather uneasy in testing and providing demarcation for the distinction of scientific knowledge from the so called non-scientific knowledge. It is not possible to have scientific thought without knowing or at least even unconsciously going through the process of philosophy, which provides complete freedom in scientific thinking. Although, today many academicians may think that they are producing scientific papers without thinking about the philosophical ingredients in their approach, in fact, their procedure has unconsciously scientific philosophical scraps. Complete freedom of philosophical thinking provides many scenarios about any phenomenon concerned, but logic eliminates tremendous amount of these on the basis of contrary results to logic or at least to common sense. Of course, common sense is unreliable much time, but it is common in all human beings for concluding or decision making about the case. Philosophers of science seek for exploration of general scientific characteristics that mostly relate to its function as a knowledge-producing activity such as the nature itself and all kinds of explanations, the nature of its validation procedures, its patterns of development, the truth-state of its theories and alike.

3.12 Science Philosophy

One of the powerful fuzziness sources in human ever-emerging desires of various kinds is from simple physical desires, which are shared with other animals to much more complicated desires specific for human nature. Every desire agitates the mind and distracts the process of concentration indispensable for an act of understanding to be productive. The stronger an emergent desire and the higher the degree of agitation it stirs up, the less the degree of concentration of mind and the less the degree of concentration, the fuzzier the process of thinking, the lower the degree of understanding. Most of the desires self-propel their intensity—the more one tries to satisfy them, the higher become the demand; the way of moderation—the 'middle way' as in the Hadith of Prophet Mohammad (pbuh), is hard to follow when the fire of desires is burning inside us and making the minds restless, turbulent and obstinate. Amidst of such feelings the human mind is completely free in thinking including every extreme. This is referred to as the philosophical thinking (Fig. 3.3), which must be filtered later through the logic rules for deducing proper, meaningful and useful statements (arguments) leading to plausible conclusions.

The restlessness and turbulence of minds are permanently intensified by the stress in which one lives due to the competitiveness inherent in today's society and the helplessness of majority of us to get out of the social boxes and cages (in which we have been pushed by economic forces too strong to resist), even if we desperately desire to. Although the strength of passion with which we pursue truth and understanding is a powerful stimulator and 'energizer' of thinking, understanding needs 'peace of mind'—a mind, which is calm and cool, composed and collected.

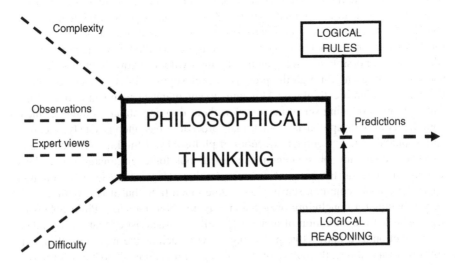

Fig. 3.3 Scientific philosophical thinking

Paradoxically enough, while being sources of fuzziness, mind and desires are, at the same time, key factors for overcoming (transcending) them, especially if they relate to problems deeply rooted in human experience. The fuzzy concepts in understanding problems that emerge out of life complexity as it unfolds cannot be resolved at the same level of knowledge that we have when these problems appear. Only when our consciousness is expanded i.e. raised to a higher level, then the tension fades away and the problems, being seen in a new light, are no longer problems. When problems dissolve, we say that the fuzziness related to them has been transcended.

The qualitative jump of consciousness to a higher level results in transcending the fuzziness. As far as consciousness is a holistic characteristic of human, and perhaps not just of human, nature and not only a product of mind, its growth and transformation are possible when the factors responsible for the integrity of all three inseparable constituents of human individuality, which are body, mind and soul, become simultaneously activated. This simultaneous activation ('firing') is referred to as a consciousness resonance and hence, "the fuzziness of understanding can be transcended when the consciousness resonance occurs."

Initial conceptualization of an event investigation should have philosophical reasoning for gaining insight about its multi-dimensional aspects. To philosophize about the nature of a problem involves assaying existing knowledge to identify variables and propose relationships between these variables (Chap. 5). Usually this involves formulating propositions that can be tested.

Philosophizing is indispensable to research processes. Philosophy—meaning literally, 'a love of wisdom' (from Archaic Greek; philosophia)—was originally a blanket term used for dealing with all questions about humanity, the physical universe and the manner in which perceptions about the two were perceived to interact through rational human thought [24]. Whilst questions about the physical universe (i.e. 'natural' philosophy) were gradually hived off into specific disciplines such as biology, chemistry, physics and so on, the present concerns of 'philosophy' have become increasingly focused in three areas. These are:

- Epistemology or enquiry into the nature and ground of experience, belief and knowledge;
- Metaphysics, or the immanent or transcendent investigation of the world and of what really exists;
- Ethics or how people should act in general, rather than as a means to an end [25].

Firstly, there is convincing evidence to suggest that human meaning, and consequently, our interpretation and understanding of words, terms and phrases changes over time [26, 27].

Verifiability of scientific knowledge or theories by logical positivists means on the classical grounds that the demarcation of science concerning a phenomenon is equal to 1 without giving room for falsification of Popper (1902–1995). The conflict between verifiability and falsifiability of scientific theories includes philosophical grounds that are fuzzy but many scientific philosophers concluded the case with Aristotelian logic of crispness which is against the nature of scientific development. Although many science philosophers tried to resolve this problem by bringing into the argument the probability and at times the possibility of the scientific knowledge demarcation and scientific development, unfortunately so far the "fuzzy philosophy of science" has not been introduced into the literature [14, 28].

These are the two terms used in statistical sense to describe any phenomenon, which is unpredictable with any degree of certainty. An illuminating definition of random is provided by famous statistician Parzen [29] as,

> A random (or chance) phenomenon is an empirical phenomenon characterized by the property that its observation under a given set of circumstances does not always lead to the same observed outcome (so that there is no deterministic regularity) but rather to different outcomes in such a way that there is a statistical regularity.

The statistical regularity implies group and subgroup behaviors of a large number of observations so that the predictions can be made for each group more accurately than for individuals. For instance, provided that a long sequence of temperature observations is available at a location, it is then possible to say confidently that the weather will be warm or cool or cold or hot tomorrow than specifying numerically by prediction the degree of centigrade. As will be explained in the later sections, the statistical regularities are reflections of astronomical, natural, environmental and social effects' combination. The global climate change discussions are based on the fossil fuel pollution in the lower atmospheric layers due to anthropogenic activities. The climate change effect is expressed by different researchers, but its intensity cannot be determined with certainty over the coming time epochs. Statistical regularity implies further complete unpredictability for single or individual events.

Deterministic phenomena are those in which outcomes of the individual events are predictable with complete certainty under any given set of circumstances, if the initial conditions are known. In the physical and astronomical sciences, traditionally deterministic nature of the phenomena is assumed. It is, therefore, necessary in the use of such approaches the validity of the assumption sets and initial conditions. In a way, with idealization concepts, assumptions and simplifications, deterministic scientific researches yield conclusions in the forms of algorithms, procedures or mathematical formulations, which should be used with caution for restrictive circumstances. The very essence of determinism is the idealization and assumptions so that uncertain phenomenon becomes graspable and conceivable to work with the available physical concepts and mathematical procedures. In a way, idealization and assumption sets render uncertain phenomenon into conceptually certain situation by trashing out the uncertainty components. A significant question that may be asked at this point is whether there is not any benefit from the deterministic approaches in the earth and atmospheric studies, where the events are uncertain? The answer is affirmative, because in spite of the simplifying assumptions and idealizations, the skeleton of the uncertain phenomenon is captured by the deterministic methods.

Historically, uncertainty and imprecision have been digested in the logical reasoning of early philosophers and researchers without any mathematical formulation. Initially only observations were the prime source of information, and accordingly, philosophers deduced their rational statements about the phenomenon under consideration. For instance, the earth was conceptualized as a flat cylinder swimming in the water (Thales, c. 624 BC–c. 546 BC). Although the real position of the earth is complicated, in early times simple observations led to simple conclusions depending on rational reasoning. In this manner, uncertainty was eliminated by rational conclusions as first human perceptions. In early ages, scientific conclusions were filtered by logical reasoning from the world of uncertainty or extreme complexity. Speculations were among the useful suggestions for the description of the natural phenomenon behavior. It is possible to summarize the early studies as having their foundations on linguistic, and hence, logical bases. This is the reason why very early instrument that helped humans to arrive at rational conclusions was the 'logic' as suggested by Aristotle (384 BC–322 BC).

In addition to observations as the sole information source, in the middle ages, measurements started to provide additional base for scientific thinking through numerical data. Measurements as a new phase of information reduced the uncertainty and ambiguity in the scientific works greatly. Coupled with the numerical data base, sole logical rule base derived conclusions were reevaluated, and hence, more refined scientific conclusions were obtained. For instance, it is found that the earth has a spherical shape instead of a flat cylinder, which kept human mind captive for many centuries. One of the Eastern Muslim philosophers AlFarabus (890–935) classified then known sciences into two categories as probabilistic and deterministic. Among the probabilistic sciences he counted physics, mathematics, geometry and logic, which app become uncertain in the nineteenth century as quantum physics, chaos mathematics, fractal geometry and fuzzy logic, respectively.

Starting from sixteen century onwards, the probability theory, calculus and mathematical formulations took over in the description of the natural real world system with uncertainty. It was assumed to follow the characteristics of random uncertainty, where the input and output variables of a system had numerical set of values with uncertain occurrences and magnitudes. This implied that the connection system of inputs to outputs was also random in behavior, i.e., the outcomes of such a system are strictly a matter of chance, and therefore, a sequence of event predictions is impossible. Not all uncertainty is random, and hence, cannot be modeled by the probability theory. At this junction, another uncertainty methodology, statistics comes into view, because a random process can be described precisely by the statistics of the long run averages, standard deviations, correlation coefficients, etc. Only numerical randomness can be described by the probability theory and statistics.

Overwhelming amount of uncertainty associated with the complex systems is nonrandom in nature. For instance, 'drought', 'flood', 'surplus', 'dry', and 'deficit', 'wet' are commonly used concepts in engineering, economy and earth sciences that include linguistic uncertainty, which becomes vaguer when one says 'severe flood', 'very dry', 'almost wet', 'intensive drought', 'slightly wet', etc. Fuzzy set theory helps to deal with this type of vagueness in modeling the economic and natural events with imprecision and/or lack of information regarding the problem at hand. In any modeling, most often one understands that there is some lack of complete information in the solution. Some of the information may be judgmental and qualitative in words. All these can be incorporated in the fuzzy logic modeling of the processes concerned in addition to perceptions about the phenomenon.

In general, there are two uncertainty types, namely, random uncertainty and nonrandom (inherent) uncertainty. For random uncertainty, the classical example is the question of "What is the probability of observing a dry year from a sequence of say, 12-year record? It is assumed that the probability of wet and dry year occurrences is equally likely, mutually exclusive and completely random (independent). Given the information that there are 4 dry and 8 wet years in sequence, the probabilities of random wet and dry year occurrences are $4/12 = 0.32$ and $8/12 = 0.78$, respectively. Hence, random uncertainty deals with events. Once the event occurs, the uncertainty goes away for that particular event.

Another type of uncertainty has been acknowledged and even measured in classical information theory long before the introduction of fuzzy logic concepts, namely, uncertainty emanating from lack of specific information regarding the object of interest. A typical example of this type of uncertainty is the one caused by the finite resolution of any measurement instrument. For example, when a measurement is accurate to one decimal digit only, it means that two or more digit decimals are indistinguishable.

A sensible way to model the uncertainty derived from the coarseness of measurement instruments is to partition the interval of real numbers representing the range of values of the discussed variable into disjoint subintervals, such that values within each subinterval would be considered indistinguishable. The subintervals are usually labeled by real numbers, which may be their respective means or other representative numbers such as median or mode; and these values that fall within the same subinterval are perceived as the same state of the variable, labeled by the same number.

On the other hand, nonrandom (inherent) uncertainty deals with characteristics of the objects themselves, and arises from our attempt to classify or categorize them. The classical question is to ask which years are dry (black) and which are wet (white)? If there are gray years of different tones then the answer to such a question becomes fuzzy, i.e. vague, ambiguous and incomplete.

In certain circumstances, both types of uncertainties may exist. For example, we may be given a sequence of observations of years of varying degrees between 'pure dry' and 'pure wet'. Then a question that may be formulated is "What is the probability that the year, the engineer considers, will conform to our concept of 'wet'? In these statements, the words 'wet' and 'dry' are vague, incomplete and ambiguous information, and hence they can be represented by fuzzy sets (Chap. 4).

Early perceptions, knowledge, information and concepts are derived from observations, experiences and occasional experiments. In the meantime, science is separated from philosophy with its own axioms, hypotheses, laws and formulations, especially, after the renaissance in the seventeenth century with deterministic (crisp) world, where uncertainty was not accounted among the scientific knowledge. However, today almost in all the branches of science, uncertainty ingredients are significant including fuzzy (linguistic, verbal) information. In engineering and earth sciences (hydrology, meteorology, geology, etc.), economics and social sciences some topics have never gone through the stage of complete determinism. With the advancement of numerical uncertainty techniques such as probability, statistics and stochastic principles scientific progresses had rapid developments quantitatively, but still leaving aside the qualitative (linguistic, intuitive) sources of knowledge and information. Famous philosophers and scientists alike started to spell out the uncertainty, and fuzzy ingredients that are essential basis of scientific progress. For instance, Russell [30] stated that,

> All traditional logic habitually assumes that precise symbols are being employed. It is, therefore, not applicable to this terrestrial life but only to an imagined celestial existence.

On the other hand, Zadeh [31] said that,

> As the complexity of a system increases, our ability to make precise and yet significant statements about its behavior diminishes until a threshold is reached beyond which precision and significance (or relevance) become almost mutually exclusive characteristics.

It is clear today that description and generation leading to satisfactory mathematical structure of any physical actuality are often unrealistic requirements. The phrase,

> So far as the laws of mathematics refer to reality, they are not certain. And so far as they are certain, they do not refer to reality.

as stated by Einstein.

3.13 Uncertainty in Effect-Cause Relationship

There has been a good deal of discussion and curiosity about the natural event occurrences during the last century (Chamberlain 1904–1978). These discussions have included comparisons between uncertainty in earth and atmospheric sciences and uncertainty in physics which has, inevitably it seems, led to the question of determinism and indeterminism in nature [32].

At the very core of scientific theories lies the notion of "cause" and "effect" relationship in an assumed absolute certainty domain in scientific studies. One of the modern philosophers of science, stated that:

> ... to give a causal explanation of a certain specific event means deducing a statement describing this event from two kinds of premises: from some universal laws, and from some singular or specific statements which we may call the specific initial conditions.

According to him there must be a very special kind of connection between the premises and the conclusions (consequents) of a causal explanation, and it must be deductive. In this manner, the conclusion follows necessarily from the premises. Prior to any mathematical formulation the premises and the conclusion consists of verbal (linguistic) statements. It is necessary to justify every step of deductive argument by citing a logical rule that implies some relationship among cause and effect variables. On the other hand, the concept of "law" lies at the heart of deductive explanation, and therefore, at the heart of the certainty of our knowledge about specific events. A simple cause and effect modeling is given in Fig. 3.4.

In general causes and effect are observed or measured, and they are known, but the relationship translator between them is not known in many engineering, scientific, economic and social events. It is, therefore, very important to be able to identify some of the features of the relationship through deductions, inferences, linguistically reasoning and perhaps at the end by well-established and convenient mathematical deterministic or uncertainty methodologies.

Recently, the scientific evolution of the methodologies has shown that the more the researchers try to clarify the boundaries of their domain of interest, the more they become blurred with other domains of research. For instance, as groundwater engineers try to model the groundwater pollution as one of the modern nuisances of humanity, so far as the water resources are concerned, they need information about the geological environment of the aquifers, meteorological and atmospheric conditions for the groundwater recharge and social and human settlement environmental issues for the pollution sources. Hence, many common philosophies, logical basic deductions, methodologies and approaches become common to different disciplines and the data processing is among the most important topics, which include the same methodologies applicable to diversity of disciplines. The way that earth, environmental and atmospheric scientist's frame their questions varies enormously but the solution algorithms may include the same or at least similar procedures.

Any natural phenomenon or its similitude occurs extensively over a region, and therefore, its recordings or observations at different locations pose some questions as, for instance, are there relationships between phenomena in various locations? In such a question, the time is considered as frozen (steady state) and the phenomenon concerned is investigated over the space and its behavioral occurrence between the locations. Answer to this question may be provided descriptively in linguistic, subjective and vague terms, which may be understood even by non-specialists to a great extent. However, their quantification necessitates objective methodologies, which are one of the purposes of the context in this book.

During the last decade several publications show the increasing significance of philosophy of engineering in various parts of the world [1, 14, 33].

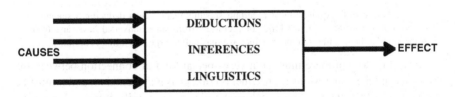

Fig. 3.4 Cause-effect relationship

As an esteem of the eastern thinking, philosophical objects may be raised by logical premises and implications along three basic mental activities, namely, imagination, conceptualization and subsequently idealization jointly leading to idea generations. Since the existence of terrestrial life human beings have interaction with nature, which has provided the basic material in the form of objects and events evolving with time and space for the human mental activity chain as shown in Fig. 3.5 [14].

At the early stages of human history or during the childhood of any individual, these stages play roles in different proportions and with experience they take final forms. Even so each one of the chain element in the thinking process includes uncertainties because imagination, conceptualization and idealization stages are rather subjective from individual to others. At any stage of human thinking evolution the premises includes, to a certain extent, uncertainty elements such as vagueness, ambiguousness, possibilities, probabilities and fuzziness. Implication of mathematical structure, from the mental thinking process point of view, might seem exact, but even today it is understood as a result of scientific development that at every stages of modeling, whether physical or mechanical, there are uncertainty ingredients, if not in macro scale, at least at micro scales. It is clear today that mathematical conceptualization and idealization leading to satisfactory mathematical structure of any physical actuality are unrealistic requirements.

At the very elementary stages of mental thinking activity, objects are thought as members or nonmembers of a given or physically plausible domain of variability. This brings into consideration sets, which include possible outcomes or basis of any phenomenon. In the mathematical and physical sciences, almost invariably and automatically these elements are considered as either completely member of the set or completely outside the same set. Hence, the Aristotelian logic of pairs in the form of 0 or 1; positive or negative; yes or no, etc., are employed at the foundation of physical phenomenon, and thereon, in its mathematical modeling. However, Lotfi Askerzade Zadeh [34] suggested instead that membership value that varies between 0 and 1, inclusive. Hence, fuzzy sets play intuitively plausible philosophical basis at every stage of the mental activity.

Among many career groups, most frequently engineering has applications based on numerical approaches through ready equations and formulations. Unfortunately, the artistic facet of engineering is forgotten, and hence, verbal ingredients are ignored completely as if engineering does not need philosophy and

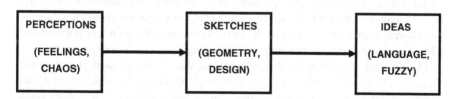

Fig. 3.5 Philosophical thinking stages

logical aspects systematically. Engineers are trained in such a way that their major aids in the career achievements are in the forms of numerical calculations. Ready use of equations and formulations may cause ignorance of philosophy (scientific philosophy), logical rule sets and rational production means. Consequent impacts of such a training push engineering students towards static and dogmatic knowledge without even any modification in the memorization process during education stages and after the graduation. Among the main reasons of such a training are insignificant attachment to linguistic aspects of knowledge and frequent usages of concrete mathematical formulations suggested by previous engineering or scientific works and their repeated descriptions to problems as prescriptions that are also crisp and concrete. Without linguistic thoughts, philosophical principles, logical inference rules and dynamic rational reasoning, even expert engineers may think that they are better than new graduates in the sense of knowledge generation and innovative productions. If asked, even expert engineers may not provide linguistic and philosophical mechanism of any methodology in detail with creative thoughts that may root future developments in engineering aspects. This is due to the fact that they use available methodologies and techniques without critical reasoning and questioning. In this manner, philosophical reasoning has been driven away from the engineering curriculum and in some countries even the same course contents are repeated each year for decades without any improvement.

It is well known by everybody that behind any equation, algorithm, procedure or solution there is linguistic and logical arena that leads to mathematical formulations. Due to the teachers' lack in linguistic explanation coupled with philosophical thinking component absence renders the educational system into a memorization institute without or with very little critical and rational reasoning. Prior to anything, the very word of engineering has in its definition description, geometry and design all of which require verbal information to reach at end aims for the benefit of a society. Any engineer, before proper planning and project should be empowered by verbal knowledge and information with reasoning and such information must be combined with his/her abilities so as to produce the best, cheap and economic solutions in the shortest time possible by deciding from various alternative solutions in an optimal manner. Unfortunately, today in engineering institutions more than any other topic, mathematics is pumped into the minds of young candidates without philosophy, logic, geometry and scientific principles. If these topics are brought into the consideration of engineers then they may be empowered to translate linguistic information and knowledge into mathematical symbols in equations and formulations. Dynamic consciousness about the philosophical and logical principles provides more effective and creative mind activities in engineers' thoughts leading to new ideas, formulations, methodologies and procedures. There is no benefit in the formulation memorization and static knowledge, but on the contrary, they may give to an engineer an inferiority complex.

Phenomenal development depends on suitable boundary and initial conditions. In the history, knowledge generation phenomenon is achieved in a sustainable manner by many civilizations, but at times meagerly in a consequent manner. In any civilization, mutual support of individuals, institutions and many

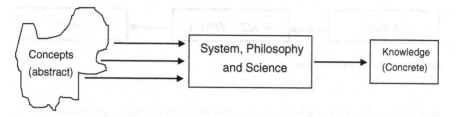

Fig. 3.6 Abstract-concrete transformations

establishments may start initially in the form of abstractions, but by time they gain concrete structures in the forms of usable knowledge and information (Fig. 3.6).

Even though initially available knowledge may not be understandable and graspable, it takes shape in human thinking by time and reaches a concrete form. Further development of such a knowledge intermingling with philosophical ingredients, and especially, scientific philosophy renders it to scientific knowledge that can be critical for further discussions. This indicates that philosophy is a knowledge generation means in mind.

At first glance, knowledge that takes scientific shape in this manner can be thought under two categories, as useful and unusual. It is necessary to define what one understands from useful knowledge. Any useful knowledge make human happy not only materialistically but also spiritually. Once this knowledge enters any society, it serves the society and generates different functions. It has two stages, as prior to and posterior to knowledge formations. The former stage provides benefits to the society whereas the latter involves philosophy and science for further critical assessment of the same knowledge. It is in the second stage where engineering concepts are efficient, positive and effective. Any society with a harmonious combination of these two stages has a basis for knowledge generation. This indicates that knowledge is like a seed that helps to generate useful sets of information.

3.14 Fuzzy Models

Classical science problem solution approaches work with crisp and organized numerical data on the basis of two-valued (white–black, on–off, yes–no, etc.) logic. Natural or social sciences have almost in every corner gray fore and back grounds with verbal information. It is a big dilemma how to deal with gray information for arriving at decisive conclusions with crisp and deterministic principles? Fuzzy logic principles with linguistically valid propositions and vague categorization provide a sound ground for the evaluation of such information. The preliminary step in fuzzy logic is the conceptualization of natural and social phenomena with uncertainties in its input, system and output variables. Such an approach is necessary not only to visualize the relationships between different variables, but

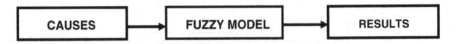

Fig. 3.7 Simple models for thinking

also to furnish a philosophical detail about the system output generation mechanism through a set of logical rules without mathematical formulations.

The preliminary step is a genuine logical and uncertain conceptualization of the phenomenon with its causal and result variables similar to Fig. 3.5 that are combined through the fuzzy logical propositions as in Fig. 3.7).

It is emphasized herein that in an innovative education system, the basic philosophy and fuzzy logic justifications in problem solving should be given linguistically prior to any crisp basis such as mathematics or numerical algorithms. In this way, the researcher will be able to develop his/her creative and analytical thinking capabilities with the support of experts who have been working on the similar aspects for many years. Since, the modern philosophy of science insists on the falsification of current scientific results, there are always room for ambiguity, vagueness, imprecision and fuzziness in any scientific research activity. Innovative education systems should lean more towards the basic scientific philosophy of the problem solving with fuzzy logical principles.

3.15 Chaos Models

Modeling through dynamic equations is available in physics considering the conservation of the mass, momentum and energy coupled with the thermodynamic principles and the state equation of gasses. Without any boundary or initial conditions differential equations are obtained in a straightforward manner with no difficulty at all. However, their solutions for practical applications pose different questions and difficulties among which are the following points.

A. The differential equations are derived under rather restrictive assumptions of uniformity, homogeneity, isotropy etc.
B. Geometrical configuration is considered as a very small cube with linear changes of the variables considered along its sides. Consideration of very small cube size in the derivations implies the assumptions of homogeneity, uniformity and isotropy automatically. Such a combination of assumptions further implies the use of the arithmetic average and linearity concepts.
C. If no boundary or initial conditions are defined and the dynamic equations obtained in this manner are general and condition free. The final shapes of the differential equations for the dynamic system can be applicable at any time and space theoretically. Practical applications bring to mind the identification of the boundary and initial conditions.

D. No element of chance is incorporated in the derivations, and therefore, they describe average behaviors of the system.
E. They are unsolvable analytically but their solutions are possible only through the numerical (finite difference, element, or boundary element techniques). Therefore, rather powerful computers and reliable solution algorithms are necessary. Furthermore, in any numerical solution technique stability condition is necessary with a definite amount of error level acceptance.
F. Although the equations are expected mechanistically to satisfy the ideal conditions under the light of a set of simplifying assumptions, their practical solutions, even though the aforementioned difficulties are avoided, still are not possible practically. This is due to the fact that, the natural phenomenon such as the climate and meteorological events is continuous in the atmosphere and their measurements are possible only at a set of irregularly scattered station locations. It is unfortunate that these locations do not coincide, say, with the nodal points of finite difference solution mash. It is necessary prior to even attempting to solve the dynamic equations to transfer record values measured at station sites to already decided mesh nodes. Any approximation in this procedure will later be reflected in the numerical solution of the dynamic equations. Since the solutions of dynamic systems by numerical techniques provide chaotic behaviors due to the initial values, what if different researchers employ different transfer methodology for the calculation of nodal values? Difference in the techniques will lead to different values at these nodal points, which will trigger the numerical solution of the dynamic system equations then each researcher is expected to have rather different solutions as a result of chaotic effects.
G. Another practical problem is after the solution of the dynamic system numerically, the predictions or estimations are available at the nodal points. This time they need transfer to the points where there are no measurements. It will introduce another source of error into the procedure.
H. Dynamic system expressions are in the forms of integral–differential equations. According to calculus, provided that the variables are continuous within a certain domain of variation along time and/or space reference systems, their substitution into the difference or integration term is calculable. Unfortunately, in practice their measurements are discontinuous and rather in the form of random behavior as time series and the differential equations do not conform to these series. As stated before since the dynamic system equations are average based on the average behaviors of the variables, it is necessary to use finite time duration averages of the time series for their match with equations. Such an approach imports into the solution algorithm statistical concepts as the variance, correlation coefficient, covariance, etc.

Most often in practice, rather than the dynamic equations of the phenomenon concerned, time series responses are available in a series of measurements. The question lies in whether it is possible to deduce the number of variables for the description of this dynamic system? Recently, phase diagrams are employed for this purpose. They are the graphs, which show the evolution of the phenomenon with time along axes of different variables.

In the case of climate modeling such series are abundant, and since atmospheric phenomena present very complex and sophisticated structure, it is not possible easily to model them through analytical and physical approaches. The remaining open door for their behavior assessment is the use of time series analysis. The classical time series analysis does not provide any insight into the dynamism of the phenomenon but only about its mechanical decomposition into various trends.

Representation of time series on one, two, three or more dimensional phase diagrams might exhibit different patterns that give impressions about the chaotic behavior of the phenomenon concerned. If, for instance, rather regular shapes are observable in the phase diagrams then one can say that the behavior of the phenomenon evolves along this shape, which is referred to as the "strange attractor". The reason for such a label is that all the phase evolutions along the time axis of the phenomenon falls on this shape for sure, but it is not certain, which point on the strange attractor corresponds to the next evolution. Hence, strangeness comes from rather unusual geometric shapes, which are not available in the conventional Euclidian concepts. These shapes cannot be quantified with Euclidian measures, but rather fractal geometric approaches are used in terms of the fractal dimensions. This means that although the phase diagrams are on Euclidian spaces, but the figure that might emerge as the strange attractor of a phenomenon cannot be accounted by the Euclidian geometry, but by the fractal geometry which expresses dimensions in decimals.

3.16 Fractal

Many researchers in the past have wondered the validity of the Euclidian geometry in modeling natural phenomena concerning different physical, scientific, technological and engineering events. For instance, Einstein in his relativity theory could not model the geometry of space with the Euclidian geometry, and therefore, he used Riemann geometry [35]. Since then many versions of Euclidian geometry have been proposed such as the Lobachevski and Riemann geometries. These versions have also many common properties such as continuity, linearity and smoothness, which are basic requirements of differential and integral calculus. The use of classical geometric shapes is not enough for every modeling and simulation. It is well known that natural objects are rough and discontinuous such as snowflakes, coastlines, fracture surfaces, clouds, mountain silhouettes, terrain topography and many other occurrences. These objects cannot be represented by Euclidian geometry except after a set of assumptions and axioms. It is interesting to notice that the theory about decimal dimension geometry is based mathematically on conception objects and the first analyses are developed about 80–100 years by Housdorff [36]. The applications in the field of material sciences are only very recent for the last three decades at the most.

The scale invariance of natural geometries is one of the first concepts that should be considered in the interpretation of any natural phenomena. Because of scale invariance, the length of a coastline increases as the length of the measuring rod decreases according to a power law, which determines the fractal dimension of the coastline. For instance, under a variety of circumstances, the frequency size distributions of various natural events are shown to have fractal dimensions. Among such events are faults, rock fractures and fragmentations, mineral deposits, oil fields and earthquakes in addition to the topography of a region as continuous variable. Mandelbrot [37] used fractal concepts to generate synthetic landscapes that look remarkably similar to actual landscapes. For instance, the fractal dimension is a measure of fracture roughness. On the other hand, a fractal distribution is the only statistical distribution that is scale invariant. An important question remains how fractal distributions are related to the governing physical laws. Furthermore, the chaotic solutions of dynamic equations obey fractal statistics in a variety of ways. Another class of models that yield fractal statistics involves the cellular automata concept.

Although the precise definition of the fractal term is elusive, fractals can easily be grasped by considering some simple geometric forms in the forms of deterministic fractals before proceeding to their natural analogues, random fractals. Triadic Koch curve is such a deterministic fractal shape, which provides fundamental concepts about the fractal geometry. Fractal geometry is currently of major interest in many fields. Fractals are concerned with probability modeling of natural phenomena size-distributions and in this regard the Pareto distribution began to replace the logarithmic normal distribution as a model for natural events. It has been shown by Mandelbrot [37] that the Pareto distribution is the probability distribution characteristics of fractals. The Pareto distribution is related to fractals, because it has a power-law form with two important scaling (self-similar) properties, namely, scaling under lower truncation and asymptotic scaling under addition.

Fractals are defined in general as objects made of similar parts to the whole in some way; either exactly the same except for scale or statistically the same. In short, fractals are self-similar or scaling, that is, invariance against changes in scale or size (scale-invariance).

References

1. Li B (2002) An introduction to philosophy of engineering. Daxiang Publishing Press, Zhengzhou
2. Koen BV (2005) An engineer's quest for universal method. Keynote address, norms, knowledge, and reasoning in technology conference, sponsored by the Philosophy and Ethics of Technology Department, Technical University of Eindhoven, Boxmeer, The Netherlands, 3–4 June
3. McMullin E (1987) "Alternative approaches to the philosophy of science", scientific knowledge. Kourany JA (eds) Wadsworth Publishing Company, Belmont
4. Quinton A (1995 The ethics of philosophical practice. In: Honderich T (ed.) The oxford companion to philosophy. Oxford University Press, Oxford, p. 666
5. Dimitrov V, Hodge B (2002) Social fuzziology. Physica-Verlag, Heidelberg
6. Şen Z (1974) Small sample properties of stationary stochastic processes and the hurst phenomenon in hydrology. Unpublished Ph. D. thesis, Imperial College of Science and Technology, University of London, p 256

7. Pitt J (2011) "Fitting Engineering into Philosophy", in the forum on philosophy, engineering and technology, philosophy, technology and engineering series, vol 3. Springer, Amsterdam (Forthcoming)
8. Goldberg DE, De Poel IV (2010) Erfahrung und Prognose. Vieweg, Braunschweig. p 5
9. De Poel IV, Goldberg DE (2010) Philosophy and engineering. An emerging agenda. Springer, The Netherands, p 361 ISBN 9789048128044
10. Davis M (2005) Thinking like an engineer. Studies in the ethics of a profession. Oxford University Press, New York and Oxford
11. Şen Z (2012) Ancient water robotics and Abou-l Iz Al-Jazari. Water science and technology: water supply. IWA specialized conference on water & wastewater, Technologies in ancient civilizations Istanbul-Turkey, 22–24 March 2012
12. Von Karman T (1911) Nachr. Ges. Wissenschaft. Göttingen Math Phys Klasse 509–517
13. Sokal RR (1974) Classification: purposes, principles, progress, prospects. Science 185(4157):1115–1123
14. Şen Z (2011) Fuzzy philosophy of science and education. Turk J Fuzzy Syst 2(2):77–98
15. Luegenbiehl HC (2010) Ethical principles for engineering in a global environment. In: De Poel IV, Goldberg DE (eds) Philosophy and engineering. An emerging agenda. Springer, Dordrecht, pp 147–160 ISBN 9789048128044
16. Mitcham C (1994) Thinking through technology The path between engineering and philosophy. University of Chicago Press, Chicago, p 397 ISBN 0226531988
17. Popper KR (1934) Logik der Forschung. Mohr, Tübingen
18. Reichenbach H (1938) Erfahrung und Prognose. Vieweg, Braunschweig
19. Goldmann S (1990) Philosophy, engineering and western culture. In: Durbin Paul T (ed) Broad and narrow interpretation of philosophy of technology. Kluwer Academic Publishers, Dordrecht, pp 125–152
20. Mitcham C, Mackey R (2010) Comparing approaches to the philosophy of engineering: including the linguistic philosophical approach. Van de Poel IV, Goldberg DE (eds) Philosophy and engineering. An emerging agenda. Springer, Dordrecht, pp 49–59
21. Vermaas PE (2010) Focusing philosophy of engineering: analysis of technical functions and beyond. Van de Poel I, Goldberg DE (eds) Philosophy and engineering. An emerging agenda. Springer, New York, pp 61–73
22. Kroes P, Meijers A (2000) Introduction: a discipline in search of its identity. In: Kroes P, Meijers A (eds) The empirical turn in the philosophy of technology. Research in philosophy and technology, vol 20. JAI Press, Amsterdam, pp xviii–xxxv
23. Merchant C (1980) The death of nature: women, ecology, and the scientific revolution. Harper and Row, San Francisco
24. Mautner T (2000) Ontology. The penguin dictionary of philosophy. Penguin, London
25. Bullock A, Trombley S (1999) The new Fontana dictionary of modern thought. HarperCollins Publishers, London
26. Chantrell G (2003) Oxford essential dictionary of world histories. Berkley Publishing Group, Berkley
27. Leung SK (2000) Language and meaning in human perspective. Janus, London
28. Carnap R (1987) "The confirmation of laws and theories", scientific knowledge. In: Kourany JA (eds) Wadsworth Publishing Company, Belmont
29. Parzen E (1960) Modern probability theory and its applications. Wiley, New York
30. Russell B (1948) Human knowledge: its scope and limits. George Allen & Unwin, London
31. Zadeh LA (1965) Fuzzy sets. Inf Control 8(3):338–353
32. Leopold LB, Langbein WB (1962) The concept of entropy in landscape evolution: U.S. Geological Survey, Professional Paper 500-A, p 20
33. Bucciarelli LL (2003) Engineering philosophy. Delft University Press, The Netherlands
34. Zadeh L (1963) Optimality and non-scalar-valued performance criteria. IEEE Trans Autom Control 8:59–60

35. Einstein A (1952) The principle of relativity: a collection of original papers on the special and general theory of relativity. Dover, New York
36. Hausdorff F (1919) Dimension und ausseres mass. Math Ann 79:157–179
37. Mandelbrot BB (1982) The fractal geometry of nature. W. H. Freeman and Co., San Francisco

Chapter 4
Logic and Engineering

4.1 General

Mind has different knowledge and information generation duties among which is the rational inference that can be derived from the philosophical arena of thoughts by reasoning towards identification of relationship between causatives and end results. The most primitive and effective means of such derivations is the logical principles. Human beings are not capable easily to identify mutual relationships rationally or even experimentally between more than two variables; one is the causative (antecedent, input) and another is the resultant (consequent, output) variables. In order to get preliminary idea about the relationship two fundamental logic questions must be asked. The first one is concerned with the proportionality and it asks "Is the relationship between the two variables (antecedent and consequent) directly or indirectly proportional?" The type of relationship can be felt intuitively after reasoning under the light of available information set by logical principles, and finally, the answer is either "yes" or "no". The final answer provides the type of relationship in the linear world.

The second question may also be asked as "Is the relationship linear or non-linear?" which might not be answered in many cases properly by logical rational thinking. In such cases experimentation either in the laboratory, in the field or observations provide the necessary type of relationship whether linear or non-linear. Throughout the science history almost in all the subjects the scientific laws are concerned very simply with the first type of question, and therefore, all the scientific laws are put or stated in the linear form although they may have non-linear behaviors. All scientific laws have linear forms; Newton, Hooke, Hubble, Ohm, Fick, Fourier laws appear in the form of linearity that expresses the relationship between two variables, one dependent and the other independent. In order to bring them into practically applicable conditions, the researchers pose assumptions or restrictions in their applications. For instance, Hooke's law, which is concerned with the material sciences, assumes that the material behavior is elastic meaning that the material is subjected to work under a range of forces will not lead to

Z. Şen, *Philosophical, Logical and Scientific Perspectives in Engineering*,
Intelligent Systems Reference Library 143, DOI: 10.1007/978-3-319-01742-6_4,
© Springer International Publishing Switzerland 2014

plastic (non-linear) behaviors. This is possible only by allowing the materials to have loads such that the loading and unloading will take place along the linear relationship between the two concerned variables, which are the stress and deformation in this case.

One significant point in all the laws and formulations in engineering is that the concerned variables are considered holistically. This is tantamount to saying that the variables are visualized as a whole without any sub-degradations. This is what happens when one asks the question of proportionality, where both variables are considered holistically. They are not considered as sub-grades of "low", "medium" and "high" or any number of sub-divisions.

This chapter will provide the bases of logical thoughts in engineering works again without any indulgent into mathematical expressions or complexities. Three types of logic will be touched herein in detail as crisp (two-value, Aristotelian), symbolic and fuzzy logic aspects. In general, this chapter combines the principle elements of logical bases and their combination with engineering functionality.

4.2 Words and Sentences

As mentioned in the previous sections words are symbols that are loaded with knowledge and information. Their epistemological aspects provide accumulation of knowledge in the form of theory of knowledge. The words are scripts in the mind loaded with meanings waiting for activation and collaboration among them through human intelligence. Their combined meaningful information can be achieved through sentences in any language. Sentences provide a collective dynamism to a set of words especially scientific sentences relate causative words with resultant ones after approximate and at times crisp reasoning.

Sentences are the means of intuitional and mental formations that reflect the visualization of different subjects and their relationships in an approximate reasoning manner including vague, imprecise and partially uncertain ingredients. They can be brought into sharper meaningful statements through logical rules, which help to explain relationships between causes and consequents through a set of logical connectivity words. If a sentence provides even imprecise relationship between causative and consecutive parts then one can derive logically clear messages by making a set of assumptions.

After 1960 there was an increasing trend of computer program (software) writing as supplements to research programs. Such software cannot be achieved without logical statements (sentences), because the preparation of flow chart prior to program needs logical sequential steps. This movement gave rise to distribution of the logical structural ideas among the engineers to solve their problems speedily through the main frame computers at that time. Many researchers and engineers alike become more aware about the logical principles in search for wider diverse applications through software and hardware verification, computational linguistics, knowledge representation, etc. In order to communicate with computers a set of formal languages has been developed, which offered a wealth of alternatives to classical languages. Formal languages translate the crisp logical sentences into software statements. The computations from the computers are in terms of symbolic logic and numerical results. In order to decide and provide a possible communication between human and computer, one should decide what the best formalism is for a given reasoning or modeling task? For this purpose, a set of questions are on the researchers' way, such as "how does one compares, how does one measure, how does one test?" These are the questions that can be answered through logical statements in engineering. However, after so many years, the logic engineering or the significance of logic in engineering affairs is not well understood or known as yet because they do not appear in the engineering curriculum. Since, logic is concerned and each piece of research needs it, it is not possible to present a ready list of "recipes" of how things work. One can learn from analyzing in detail a particularly interesting case. If one is careful about the logical statements in his/her research then s/he gains experience with time, which may lead him/her to be an expert in the area of concern. Today experts are those who know their jobs linguistically by a set of relevant logical sentences.

4.3 Rational Reasoning

As already mentioned in Chap. 3 about philosophy, rationality is the characteristic of any action, belief, or desire that makes their choice a necessity. It is a normative concept of reasoning in the sense that rational people can derive conclusions in a consistent way given the information to their disposal. It refers to the conformity of one's beliefs with one's reasons to believe, or with one's actions with one's reasons for action. However, the term "rationality" tends to be used differently in different disciplines, including specialized discussions of science, economics, sociology, psychology, and politics. A rational decision is one that is not just reasoned, but it is also optimal for achieving a goal or solving a problem.

On the other hand, the word "rational" may convey information such as mental, plausible, logically acceptable, simple, short and crisp statements that may provide insight into a complex phenomenon. Rational thinking is necessary for developing claims about reality and it involves the use of logic and then science for determining the truth about reality. Checking whether a given logical sentence is wrong is a crucial part of the rational thinking process, which means that one should think critically. Rational thinking can be considered as synonymous to critical thinking.

Rational thinking capability of someone does not mean that s/he has not emotions and intuitional feelings, which are real and genuine parts of rational individuals. Most often individuals have a strong desire to maintain certain good feelings they are comfortable with. Hence, belief affects emotions and intuitions. For instance, one believes in reality that smoking is health damage, but still they may continue to smoking. This means that although they have rational thinking capacity, they may not apply rational rules in their lives. However, such subjective believes and applications are not valid in scientific logical thinking in arriving to rational conclusions.

4.4 Propositions and Rule Base

Philosophical reasoning and especially logical inferences need propositions, which refer to relational content with meaning for further rational thinking possibilities and they are declarative in the forms of IF-THEN statement implications. Each logical statement is a proposition with its antecedent and consequent parts that imply IF (antecedent) THEN (consequent) sort of relationships.

The relational content leads one to ponder whether it is true or false; or partially true or false. No need to say that the content of any proposition is subject to philosophical debate for possibly new idea generations, and hence, the overall meaning of a proposition becomes clearer after each discussion. Controversy is an imbedded property that may exist in any proposition with imprecise and vague information content. Only a set of assumptions cuts the way of controversy, because any proposition coupled with assumptions is regarded as absolute truth, which is never the case especially in scientific and engineering disciplines. The words statement and sentence are also used in the same sense of proposition subject to interpretations. Belief, desire, sense, etc. based configurations also lead to propositions, which are for exposed critical view for more general acceptance by folks. Hence, they can be interpreted as mental content of attitudes.

Propositions have also formal logic as objects of a formal language. The form of a proposition depends on the type of logic. The elements of such language are either variables (cause-reason-input or effect-result-output), predictive relationships, mathematical symbols and operators, functions, quantifiers and constants. One can propose his/her view of point about an event and then s/he must defend it against the critical debates of others.

Concepts that are obtained through rational inference can be brought together to establish meaningful sentences for description of a phenomenon. Each sentence cannot be regarded as proposition. For a sentence to be proposition there must be a syllogism in its structure. Propositions lead human to thinking and consequently to identify that the proposition is "true" or "false" in reaching a decree. Propositions can be either true or false, which is the result of two-valued (crisp, Aristotelian) logic. For example, the proposition: *Fluids, when overheated, evaporate* includes a decree as the evaporation at the end. Another proposition is: *If student works, s/he succeeds*, where "success" or "failure" are two possible outcomes. These are simple propositions, where there is an antecedent part (between IF and THEN) as object and a consequent part (after THEN) as decision or decree. If there are more than one object in the antecedent part, then they are referred to as composite propositions. Their joint effects can be expressed with one of the logical connectives, which are "AND", "OR" and "NOT". For instance: *IF temperature decreases AND snow falls AND universities are closed THEN there will be no examination*, is a compound proposition. Here, three antecedents are combined with AND logical connective. There is always quantitative or qualitative relationship between the antecedent and consequent parts, and hence, one can make decision based on a given proposition.

A set of propositions describing a phenomenon is sufficient to model it linguistically. Such a set of logical propositions is called as "rule base", where many properties of the phenomenon subject to modeling brought together based on philosophical and logical principles only. Any modeling can be regarded as an algorithm, which provides solution under the light of a rule base. Accordingly, for any modeling, the advice in this book is that the engineer should try and identify the rule base verbally (philosophical and logical principles) prior to any formulation for numerical solution. In any rule base there may be many simple or compound propositions.

It may not be possible to represent the behavioral performance of some phenomenon with a single proposition, and a set of partially mutually inclusive propositions may suffice for the presentation at an acceptance level. The collection of descriptive propositions is referred to as the rule base, because their simultaneous considerations explain the whole phenomenal behavior.

Any rule base should have IF-THEN implications in its structure and they provide the generation pattern of the phenomenon leading to inference engine search for the patterns embedded in the rules in such a manner that they match the overall qualitative and quantitative validities. In a rule base IF means when the condition is true, THEN means take action. In many cases, there is also another word as ELSE after IF-THEN, which means that when the condition (rule) is not true take another action (rule). In the logical connectives ELSE corresponds to "OR", and hence to "ORing" action. Some of the examples are as follows.

Rule base may be regarded as a mechanism which translates the combined effects of reasons into a consequent and such a mechanism helps to make inferences about the performance of the phenomenon examination for search towards beneficial conclusions.

4.5 Logical Inference

One may have different opinions, and hence, motivations for practical and theoretical reasoning. Experience helps to add more knowledge and information for deduction of rational and logical inferences that can seem more secure and if one knows risk-free inferences, and then s/he may be more alerted to the points at which there are small risk errors. Claims about inference are also intimately related to the nature of thought, to language. In most cases it is possible that ordinary language "disguises" the underlying structure of thought. Certain inferences take their security from the basic logical forms. The conceptions evolve with conceptions of first language and then logic.

Zadeh [15] distinguished two fuzzy logic directions, the first one is in the broad sense (older, better known, heavily applied but not asking deep logical questions), which serves mainly as apparatus for fuzzy control analysis of vagueness in natural language and several other application domains. Tolerance and impreciseness (vagueness) are embedded in this branch and they provide economic, quick, simple and sufficiently good solutions.

The second direction is the fuzzy logic in the narrow sense, which relies upon the symbolic logic with a comparative notion of truth developed fully in the spirit of classical logic (syntax, semantics, axiomatic, truth-preserving deduction, completeness, etc. both propositional and predicate logic). Similar to many-valued logic it is logically based on the paradigm of inference under uncertainty and vagueness. In both directions one can simply arrive at logical conclusions in terms of rule bases, which is composed of a set of rules each of which explains the phenomenon or event partially.

Logical conclusions may be derived after the act of inference based on a set of propositions that are valid for the generation mechanism of the phenomenon concerned. The output of inference is a set of logical conclusions, which are also subject to further assessment with reality either in the forms of observations, measurements or records. If the conclusions are drawn from a set of multiple observations then it is referred to as inductive reasoning inference. The final conclusion may be either absolutely correct or false, but in natural and engineering works as well as in science there is never absolute truth, and hence, the conclusions are either probabilistic numerically or fuzzy linguistically. Inferences in the forms of conclusions are based on factual information but after the conclusive ends the conclusions must be tested with further additional data or observations. There are statistical inference systems including Bayesian approaches and recently expert systems are all based on logical propositions and subsequent inference for conclusive ends in applications. In the past, engineering works (buildings, bridges, roads, etc.) are based on belief, emotion, sense and mental thinking productions all of which took the final shape after many trial and error iterations through observations and/or measurements or experience by making mistakes.

The most important property of any proposition as distinction from concepts or terms is that it can be true or false. In any proposition, the reasoning that appears due to the relationship between the antecedent and consequent parts can be true or false. Hence, according to two-value (crisp) logic one has to make a decision inferred from propositions. For decisions one must think rationally with arguments for any inference. In making arguments, and subsequent inferences, there are four ways as deduction, induction, analogy and hybridization (Chap. 2).

For deductive argument inferences the antecedent part of the proposition should include general knowledge or a set of knowledge. This means that knowledge in the antecedent part must be very general as well as extensive and consequent part must be included in it. The formal deductive argument has at least two antecedent parts (general and particular) each with relationship with each other and then the consequent part (inference) is the inference based on the antecedent contents. A number of syllogisms have been defined by Old Greek philosophers with three-part inference that can be used as building blocks for more complex reasoning. Between two antecedents and consequent there is always the word "therefore" leading to inference. For example,

	Metals expand under heat	(General antecedent)
	Cupper is a metal	(Particular antecedent)
Therefore,		
	Cupper expands under heat	(Inference, consequent)

Here the general antecedent covers the particular suggestion, and therefore, it is referred to deduction argument. In case of correct general antecedent, the inference is also correct. In any scientific work induction arguments are used most often than any other type of arguments. This is due to the collection of information pieces through synthesis they lead to more general inference and bigger picture. Thus small pieces of information and knowledge lead to final and general inference. After the inductive inference, it is possible to use it as a general antecedent and make deductive inferences conveniently.

After the establishment of a model its use leads to detailed investigations. On the other hand, after taking some samples from the environment around us, we can try to deduce some results about the general behavior of the phenomenon. Those who work in laboratory under restrictive conditions also obtain pieces of information and try to reach a general conclusion. Thus, inductive argument methodology is bound to be used more frequently in scientific studies. An example to this is the question of whether the metals expand after heating. For this purpose, each metal can be subjected to heat and one can observe whether it expands? If one tests each material s/he can reach to the following set of information pieces leading to a consequent through inference.

	Cupper expands in heating	(experimental result)
	Iron expands in heating	(experimental result)
	Zinc expands in heating	(experimental result)
	"	"
	"	"
	Aluminum expands in heating	"
Therefore,		
	Metals expand in heating	(Inference)

If this inference is accepted as true then it can be used as an input (general antecedent) for deductive argument and all other metals are covered under this inference.

Another argument is analogy which has its foundations in the similarity of different events or phenomenon (Chap. 2). In fact, analogy is a special case of induction where from two or more similar events inference can be obtained. For example,

	Ozone and oxygen materials are the same	(Proposition for analogy)
	Oxygen causes burning	(Similarity material)
Therefore,		
	Ozone causes burning	(Inference)

Here oxygen and ozone are made of the same materials, and therefore, they are similar to each other. This similarity does not guarantee that there may be a second order similarity between the two. Analogy argument is used frequently in social sciences, natural sciences and daily speeches.

Any software code cannot be without a logical rule base, otherwise one cannot write a productive computer program.

Inference engine system processes available causative information to reach effective results by employing systematic inference steps similar to a human brain. A series of inference steps are taken if there is a problem for solution involving logical principles rather than skills. Among the inference steps are deduction, association, recognition, and decision making.

4.6 Definition of Logic

Logic is a simple word but its definition is not possible without discussions and one cannot reach to a final crisp definition. There are significant differences even today between experts on this discipline. Many do not even attempt to provide a definition, and therefore, the definition of logic remains vague in the literature. Different people have given different definitions for logic. Chronologically, simple definitions of logic are arranged in approximate order as follows with relevant literature sources.

- The tool for distinguishing between the true and the false (Averroes 1126–1198).
- The science of reasoning, teaching the way of investigating unknown truth in connection with a thesis [4].
- The art whose function is to direct the reason lest it err in the manner of inferring or knowing [5].
- The art of conducting reason well in knowing things (Antoine Arnauld 1616–1698).
- The right use of reason in the inquiry after truth (Watts 1725).
- The Science, as well as the Art, of reasoning (Whately 1826).
- The science of the operations of the understanding which are subservient to the estimation of evidence (Mill 1904).
- The science of the laws of discursive thought (McCosh 1870).
- The science of the most general laws of truth (Frege 1897).
- The science which directs the operations of the mind in the attainment of truth (Joyce 1908).
- The branch of philosophy concerned with analyzing the patterns of reasoning by which a conclusion is drawn from a set of premises (Collins English Dictionary)
- The formal systematic study of the principles of valid inference and correct reasoning (Penguin Encyclopedia).

Logic and engineering for our present purposes are to be construed liberally. Engineering is about getting things done, generally building things, which realize some preconceived purpose. Logic is the sphere of formal a priori truth, encompassing mathematics, and crucially for engineering, all that supports the construction and exploitation of abstract or mathematical models. Engineering is conceived as a discipline which is to be increasingly dominated by modeling techniques permitting to the construction and evaluation of a design prior to physical fabrication of its implementation. The increasingly dominant intellectual content of engineering problem solving, the business of modeling, is at the bottom requires pure logic. Software supporting these intellectual activities can be more effective when it is built on solid logical foundations.

This prospective future development may be related to the digital revolution which we are all now expecting or experiencing. The logical revolution, as yet scarcely anticipated, flows from the same underlying imperatives about the way in which information must be represented, if we are to be able to manipulate it effectively (http://www.rbjones.com/rbjpub/logic/engl001.htm).

Digitization is a prerequisite of information being processed by computers. What can then be done with the information by computers depends on how the information is represented.

A static image represented as a bit map can be displayed but can be manipulated less effectively than a representation of the data, which contains more structural information. A movie of a dynamic three dimensional experience can be represented as a sequence of bitmaps, but to permit interactive navigation more sophisticated representations are required. Ultimately computers are necessary to understand the data, which they manipulate and to be able to reason about the behavior as well as the appearance of the system described. To represent a system in a manner, which is adequate for the purposes of many different kinds of software may be required to work with it. It is open-ended in terms of the functionality, which may be beneficially delivered based on a logical approach (http://www.rbjones.com/rbjpub/logic/engl005.htm).

Logicisation is a natural stage in representing information in ways, which permit open exploitation and manipulation. What may now be thought an exotic and improbable development will in due course be recognized as an economic imperative, not only for engineering purposes but also in education and entertainment systems, where models are equally ubiquitous (http://www.rbjones.com/rbjpub/logic/engl001.htm).

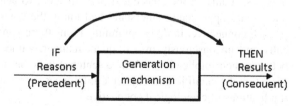

Fig. 4.1 Proposition (premise) model

In general, logic helps to deduce inferences about any phenomenon with natural languages and the end product is in the form of linguistic information and knowledge. Hence, it is very basis confirms with the fuzzy logic as will be explained later in this chapter. The development of logic and its principles have the origins in several ancient civilizations, including ancient India, China, Greece and Islam. Although it was established as a discipline by Aristotle, who established its crisp fundamentals as two-valued logic, but later Muslim philosopher Averroes defined logic as "the tool for distinguishing between the true and the false."

Generally, logic may imply two reasoning pattern in inductive (analysis), and deductive (synthesis) forms, which takes an object and examines its component parts in detail. Inductive reasoning draws general conclusions from given specific examples as a whole. The deductive reasoning draws conclusions from definitions and axioms, which considers how parts can be combined to form a whole (Chap. 2). Logical sentences are referred to as premises (propositions), which include, in general, IF-THEN structure explicitly or implicitly. The part between IF and THEN is the antecedent (inputs, reasons, causes) and the part after THEN is the consequent (outputs, results, responses). As mentioned before any black-box or grey-box modeling system has such a logical structure. Figure 4.1 shows the structure of logical statements similar to black- or grey-box models.

Figure 4.1 indicates that the propositions identify reason-result types of relationships without indulging into the internal generation mechanism of the study phenomenon. Thus, philosophy of the internal generation mechanism is not considered in detail. However, naturally such logical statements provide some information about the generation mechanism of the model. Here, a deductive modeling system is followed depending on the inferences from the wholeness of the phenomenon. Hence, rational reasoning is employed in writing down the logical rules. Any logical statement (proposition) should include in the antecedent part all the input variables with logical connectives and the consequent part including output variable should be in logical harmony with the antecedent part. There are three logical connectives, "AND", "OR" and "NOT".

4.6.1 Logical Connectives

After deciding about the inputs and outputs of phenomenon logical statements (rule bases) can be written down by convenient cooperation of logical connectives, especially in the antecedent part. In order to achieve successful rule bases, one should appreciate, understand and know the common language implications of each connective. In daily communications these connectives are used automatically and unconsciously in a frequent manner, without instantaneous distinction between their implications within a sentence, whether it is logical or not. However, for rational, scientific and objective inferences, one should know distinctions and implications of each logical connective.

When one says "AND" then two things or concepts are connected in a simultaneous occurrence manner. For example,

Ali AND Jacob are here

is a simple logical connective example. It is obvious that "AND" implies two sides and each side must exist at the same time in the same location. For the correctness of this statement, both Ali and Jacob must attend at the same time and if one of them is available only then this statement is not valid.

Another form of logical connectivity is "OR", which means that if any one of the two sides is available then the statement is valid but also if they both exist at the same time then it is also valid. For example,

when I am hungry I eat bread OR rice

is such a logical connective proposition that there is an alternative solution for the hungriness either in the form of bread or rice or both together. One should notice that in each one of the above connective statements both sides are crisp in the sense that they do not have any adjective attached to causative or consequent parts. Instead, if there are adjectives, say "little bread", "some rise", etc., then the sides are not crisp, but have uncertainty in verbal terms, which is then referred to as fuzzy uncertainty that will be explained later in this chapter.

Another important logical connective is "NOT" and for example

Horse is NOT white

statement implies that the horse is in any color except white. It may be brown, black, almost brown, partially red, with yellow patches, etc., but not white.

In order to set up logical statements (rules) about any engineering problem, one should be able to use properly logical connectives between input variables. For linguistic knowledge and logical rule establishments engineers should deepen in the meanings of logical connectives at even epistemological level. In this manner, engineers become more eligible in setting up the formulations by means of philosophical thinking, expert views and knowledge production. For example, in concrete preparation (civil engineering), deductive thoughts lead to the following preposition.

IF water AND cement AND sand AND pebble are mixed THEN concrete is
manufactured

This is a logical rule (proposition, premise) and all the inputs in the precedence part are connected by "AND", which is referred to as "ANDing" procedure. The reader should question usage of "ANDing" and "ORing" procedures.

On the other hand, in transportation from Istanbul to London the following logical statement can be written:

IF travel starts from Istanbul through Athens AND Rome AND Paris cities THEN one can
reach London.

The road to London from Istanbul has been given in a single logical statement, which is one of the possible routes, but there are several others. Now, different

Fig. 4.2 Gray model

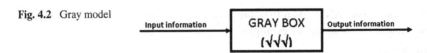

alternatives can be brought together by using "ORing" procedure, which provides within the logical statement alternative selections as follows:

> IF (travel starts from Istanbul through Athens AND Rome AND Paris cities) OR (travel starts from Istanbul through Belgrade AND Munich AND Amsterdam AND Paris) THEN one can reach London.

"ORing" provides a wide spectrum of alternative solution integration within the same model. Hence, it is possible to collect whole alternatives in a single logical statement. However, among different alternatives selection of the most suitable one is an engineering task. In order to decide on such a single outcome as decision, engineer should have some preliminary conditional knowledge. For instance, among such conditions the selection of "the shortest root" may help to reach the final decision. According to variation of conditions one of the "ORing" attachments can be valid as a final decision.

4.7 Classical and Symbolic Logic

Two-valued logic classifies all objects into two mutually exclusive categories as black-white, plus-minus, zero-one, human-not human and good-bad. There is no room between these two extremes, i.e. middle category is excluded, which is referred to as "exclusion of the middle" in classical logic studies. Since there are two categories, the inferences from this logical system cannot suit natural events, phenomenon or objects exactly. Exclusion of the middle positions leads to the acceptance of approximation right from the beginning of the affairs. In two-valued logic, truth (false) is represented by 1 (0) as numbers. Classical logic is based on crisp determinism without any room for uncertainty.

Classical black box models (classical logic) in engineering modeling are being replaced by gray box (fuzzy logic) models recently, where the internal logical rules are sought rather than crisp formulations (see Fig. 4.2).

Common sense dictates that some form of empiricism is essential to make sense of the world. In traditional quantitative educational training, the classical dualism as the tension between subjectivity and objectivity is often addressed by adopting an objectivist, empiricist or positivistic approach, and then by applying a scientific research design. Even based on classical logic, scientific thinking starts in an entirely subjective medium. Subjective thinking penetrates objectivity domain by time through imagination and visualization, and hence, there is not a crisp line between subjectivity and objectivity. Empirical works, which are based on either observations or measurements as experimental information, help to decrease the degree of subjectivity at the benefit of objectivity degree. In a way, none of the scientific formulations obtained up to now is completely crisp, but they are regarded as crisp information provided that the fundamental assumptions such as mutually exclusiveness and exhaustiveness are taken into consideration. The crispness of any scientific information can be shacked by modifying one of the basic assumptions. This implies that all the scientific principles are not crisp completely, but include vagueness, incompleteness and uncertainty even to a small extend, and hence, they can be considered as fuzzy by nature or by human understanding.

In everyday life human beings make many predictions especially on the basis of qualitative data and past experiences. Additionally, expert opinions help to shape and to refine such predictions besides the mutual discussion and confidence. In predictions there are similarities, which are the input information about the phenomenon concerned, output clues and the logical connectivity between these two sources of information. On the basis of certain clues, it is possible to make judgments about output information. The default of these judgments is the commonly available scientific thinking and its sublime version of logic (classical or fuzzy) leading to rational results. This provides ability for any individual to develop actuarial models for various real-life prediction problems.

It is possible to make predictions either by classical logic mathematical models or fuzzy logic expert views. A basic question is "Are the predictions of human experts more reliable than the predictions of mathematical models?" Experts make their predictions on the basis of the same evidence as for the mathematical foundations, but additionally they consider the usefulness of the linguistic data in the form of vague statements for the adjustment of the final model. Such vague information cannot be digested by classical logic mathematical models, because any sort of uncertainty is defuzzified into crisp numerical forms. It is, therefore, expected that the fuzzy modeling by experts considering vague information is more successful than mathematical models, which are valid for ideal cases under the validity of a set of assumptions. Among the most important problems are natural phenomena predictions, because they have the following properties.

- Even the best mathematical models have not complete reliability;
- The best results seem reasonable predictions, but somewhat unsafe, and therefore, an interval of confidence is necessary.

Similar principles are also valid for any engineering phenomena. In order to penetrate into deeper understanding and broader grasp of complexity, the emergent meanings need to be neither stable nor unstable, that is stable enough to rely upon them when generating hypotheses, concepts, and emotional attitudes, and unstable enough not to allow these concepts and attitudes to harden and become dogmas and addictions. After engineering thinking, meanings need to be fuzzy (flexible), ready to immediately respond to the changes continuously occurring in each of the numerous dimensions of reality [1].

4.8 Reasoning Principles

Logic can be defined correct and systematic thought product [8]. In this definition, there are three important conceptions "logic", "correct" and "systematic" For
a logical issue first of all it is necessary to have a topic, event or phenomenon and
thoughts based upon these and also communication to other individuals through
a "language". Logic requires communication and especially a language and for a
proper conservation one needs to abide by the grammatical rules of the language,
which is a set of linguistic rules. Grammar rules are useful not theoretical but in
practical usages. If logic is thought without the grammar, then there will not be
beneficial conclusions, but various jargons and problems. In daily speech, we do
not care about grammatical rules, because they are automatically in our memory.
In any scientific work or modeling such automations cannot be valid. One must
reason knowledge through thoughts and draw their meanings into the speech and
then ponder upon various interpretations and possible relationships deeply. One
should keep in mind that in science there are not difficulties. Simple logic and critical reasoning led to the scientific levels.

Especially, in scientific knowledge productions one has to keep in mind concepts, terms and definitions. One has to construct sentences of common sense,
logic and rationality. The sentences are similar to daily speech sentences, but they
have some discrimination. Scientific sentences must include an antecedent part
with knowledge and there must also be a consequent part for inference based on
the antecedents. If one grasps the antecedent part properly then s/he can conclude
with meaningful and rational logical outcomes. Accordingly, in any speech, paper,
book or report not all the sentences imply propositional structure, and hence,
ready for scientific inferences. For better understanding, at this point let us give an
example as,

> whoever reads this book with understanding, interpretation and location in the memory
> can make proper modeling,

this is a proposition. In this sentence as condition "reading", "interpretation" and
"location in the memory" are among the antecedents and they have knowledge
contents. Based on these antecedents, in order to reach a decision, one must infer
from the proposition a consequent, which is a kind of modeling and model output.
This proposition is either true or false, which are the two outcomes of the classical two-valued logic suggested first time by early Greek philosophers, Aristotle
before Christ.

For advancement in scientific studies propositions must be identified with
sensitivity. Many researchers make propositions about their studies and then try
to verify these in rational or experimental ways. For instance, *when metals are
heated they expand* is a proposition, which has been suggested after observation
on several metals and then generalized to cover many others. It will remain true
until someone with evidence falsifies this proposition. Up today no experimental
evidence has been found for falsification and it is subject to falsification all the
time. Some propositions can be set forth by rational or experimental suggestions

or hypothesis. It is advised in this book to start first with philosophical, rational and logical principles and then if necessary with experiments. There are cases that one cannot experiment, but rationally by observations the propositions find their existence. For example,

sun rises from the east everyday

is a proposition. However, philosophically it does not mean that tomorrow it will rise from the east.

Up to about 40–50 years ago many classical education systems considered that propositions are either true on the basis of valid or invalid antecedents. In short, this logic is two-valued and classifies all objects into two mutually exclusive classes as black-white, plus-minus, zero-one, human-not human, good-bad, etc. There is no place between these two extremes, i.e. middle classes are excluded, which is referred to as "exclusion of the middle" in logic studies. Since there are two classes, the inferences from this logical system cannot suit natural events, phenomena or objects exactly. Exclusion of the middle positions leads to the acceptance of approximation right from the beginning of the affairs. In two-valued logic truth (falseness) is represented by 1 (0) as numbers. Classical logic is based on crisp determinism without any room for uncertainty.

4.9 Symbolic Logic

There is also another logic that works through the symbols as representatives of words, terms, concepts and sentences. It is symbolic logic, which covers all the formulations and equations in engineering. This is similar to translation from one language to another. For instance, in Turkish "ağaç" represents "tree" in English. For someone with English (Turkish) native language "ağaç" ("tree") seems as a symbol. Thus, there are changes in the symbolism and shapes but the logical rational, philosophical contents remain the same irrespective of any language. In order to have a language that satisfies everybody in different language groups, symbolic logic helps more than any other logic. All the logical, philosophical and rational meanings are loaded on symbols that can be perceived by different people in their own native languages. There are three basic logical words that constitute the structure of any logic and proposition, which are "AND", "OR" and "NOT". Even these logical connectives can be symbolized, and finally, propositions take the form of mathematical symbols. Symbolic logic does not have difference from the classical counterpart as for the two outcomes of the proposition as true or false. So far in engineering education system, and thereafter graduation, all formulations and equations are based on two-value classical and symbolic logics. All thought contents in engineering are in the form of symbolic expressions as equations or formulations. Since they all depend on two-valued logic, they are all approximations for the subject of concern.

Engineering formulations today do not reveal linguistic information in terms of propositions but their implicit forms as formulas symbolically. Any engineer confronted with such symbolic expressions may not understand or even the teacher may not be able to give the philosophical, logical and rational background contents but overall meaning of each symbol in the expression. Engineers are eager to make numerical calculations by using these symbolic expressions, and therefore, they are known as "calculation" men, who care about numbers rather than verbal expressions. The philosophy of engineering content is full of linguistic expressions, which help to interpret symbolic formulations with logic in a rational manner. Whoever is empowered with the philosophy of engineering is capable to write down symbolic formulations easily, but the reverse is not valid. Unfortunately, today rather than philosophy of engineering the transfer of formulations with shallow explanations are thought in engineering education systems. This leads engineers to believe without thought that any formulation they take during education cannot be arguable or criticizeable and they can be used under any circumstance. Engineers should be able to design their thoughts without involving ready equations or if they have such equations then they should try and criticize, interpret and argue about their contents and variability at the time of practical use. Computer software programs although work on symbolic logic, their linguistic bases are put down by men. Otherwise, without linguistic background, symbolic logic does not help to solve problems. This is one of the main reasons that any engineer should have acquaintance with the philosophy of engineering principles. For instance, in a computer software the following equation must be introduces with symbols,

$$F = m * a$$

where * means multiplication and computer executes it provided that m and a are given numerically. Likewise, an engineer who takes this formulation can make numerical calculation, which means that s/he is really an engineer with rational ambitions. Unfortunately, in many engineering institutions all over the world, rather than linguistic contents only, numerical backgrounds are given, and therefore, engineers become addicted to symbolic logic coupled with classical two-valued logic training. The verbal implication of the above formulation says that

force is in direct proportionality with acceleration provided that the mass is constant.

Even though the general linguistic structure of the propositions remains the same their symbolic counterparts provide shorter expressions. For example, antecedent part of a proposition can be symbolized as A and consequent part as B, and hence, a simple proposition in symbolic logic can be expressed as,

IF A THEN B

If there are many sub-antecedents as A, B, C and D with consequent E then it takes the following form by use of "ANDing" connections,

If A AND B AND C AND D THEN E

This is very convenient for computer programming. Engineers attach extreme importance to the symbolic logic to the extent that even teachers have abstract

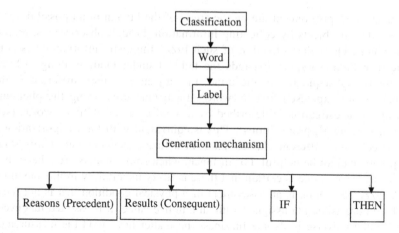

Fig. 4.3 Fuzzy concepts

ideas about linguistic principles and submit their lectures in symbolic expressions with very limited and almost static interpretations. In this way, engineers similar to computers may execute numerical calculations without suitable interpretations, criticisms or discussions and they think the satisfaction of the work ends with numerical calculation procedures. An engineer with the philosophy of engineering will not accept such an approach and will be more productive in his/her works. This engineer can bring some simple though solutions to any problem in his/her life without blindly depending on ready formulations because the philosophy of engineering principles with logical inferences and rational reasoning provide in front of the engineer much material in terms of dynamic knowledge usages. S/he first cares for linguistic information with interpretations and arguments towards a proper decision making.

Both classical and symbolic logic systems do not care for uncertainty, and their inferences are all certain, which gives the impression to engineers that they are also correct and valid universally. However, since four decades Lotfi Asker Zadeh [12] proposed a logic with uncertainty ingredients that is the fuzzy logic, which is bound to be used more frequently in future.

4.10 Fuzzy Logic Principles

There are no isolated phenomena and processes for scientific assessment, and any knowledge include uncertainty. In human perceptions, uncertainties are in linguistic forms at the imagination and description stages and such uncertainties are referred to as fuzzy. The fuzziness opens ways for changes, evolution, growth, and continuous scientific developments. Figure 4.3 gives the document of steps in fuzzy thinking for problem solving [9].

Fuzzy concepts concentrate on the study of the human mind possibilities to know external objects by collecting information through observations, records or readings. Once the collection of such fuzzy linguistic information is complete then human inquiry expands the field of understanding along different directions. Subsequently, in the mind, the objects and their different visible properties are expressed first by words. Each item concerning the phenomenon under investigation is described by a word or a set of fuzzy words (sentences, statements, propositions). This is equivalent with the categorization of the objects into different classes again in a fuzzy manner; and at this stage, crisp logic cannot be helpful. For instance, when some objects are labeled by a word, say "river", one is certain that there is fuzzy uncertainty in this labeling.

A fuzzy perception is an assessment of a physical condition that is not measured with precision, but assigned with an intuitive value. It is asserted that everything in the universe has some fuzziness, no matter how good the measurement equipment is. By using meaningful words to name the fuzzy description, the construction of engineering process is easy to understand and can be built up intuitively.

Humans are very good at recognizing by sense organs what they feel, but computers are better at counting and measuring. Fuzzy logic is very helpful in guiding the computers to find the right thing to measure and calculate. Real-world attributes are known by human perceptions through quality and quantity appreciations linguistically and/or by measurements. Different questions may be asked about individual or joint behaviors of these attributes. Humans continue to inquire knowledge by perceptions, which is a never-ending process. Fuzziness is a paramount characteristic of human perception that challenges humanity and propels the search for truth and understanding the secrets of reality. The fuzziness in human perception reveals ways of transcending it, and thus expanding the field of the human inquiry.

Fuzzy impressions and conceptions are generated by human mind, and it divides the seeable global, environmental or engineering reality into fragments and categories, which are fundamental ingredients in classification, analysis and deduction of conclusions after labeling each fragment by a word such as "name", "noun" or "adjective". The initial labeling by words is without interrelation between various categories. These words have very little to do with the wholeness of reality. Hence, common linguistic words help to imagine the same or very similar objects in our minds in a fuzzy manner. Every act of holistic understanding is inevitably fuzzy. Fuzziness and truth are not mutually exclusive, as is assumed in the crisp logic, but they do go hand in hand in every aspect of scientific research.

When consciously directing one's attention towards an external object, the object enters into the realm of one's fuzzy perception, which includes feeling, thinking, and understanding, experiencing, knowing, and applying. The levels of fuzziness correspond to the levels of one's capacity for understanding and deepening the levels of consciousness.

Measured non-fuzzy data is one of the primary inputs for the fuzzy logic models. Examples are temperature measurements by thermometers, rainfall by raingages, groundwater levels by sounders, etc. Additionally, humans with their fuzzy perceptions could also provide inputs with linguistic statements.

Recently, logical propositions and inferences become more linguistic, and accordingly, the role of the fuzzy logic increases in an unsteady manner with a variety of applications in every engineering aspect. This is in a way hand in hand with the principles of the philosophy of engineering. Fuzzy logic depends on prior to anything verbal statements and hence propositions, concepts, terms, definitions, rational relationships between inputs (reasons) as well output variables (results) in any argument. The followings are among the differences from other logic systems.

- Fuzzy logic takes into account even verbal uncertainties, which cannot be accounted by classical uncertainty methodologies (probability, statistics, stochastic, etc.). In the classical logic when one says "beautiful" its opposite is "ugly", which are understood and perceived on certainty basis. In the fuzzy logic in addition to these two classes, there are also middle classes as "more beautiful", "more or less beautiful", "rather beautiful" and "middle beautiful", and "little ugly", "middle ugly", "very ugly", etc. each one of these classes include uncertainty.
- Fuzzy logic is multitude logic. Instead of white and black only, it has grey tones in between (see Fig. 4.2). In the use of other logic systems, for instance, in the symbolic logic, all what have been thought by first inventors have been put into symbolic expression forms and engineers depend on such mathematical equations more than necessary. In fuzzy logic, they cannot do so because each problem necessitates its logical rule set forth for the solution. There are no symbolic formulations and all the solution steps and tasks are in terms of logical statements verbally. Thus, rather than mathematical equations a set of rules, rule base, takes over.

In order to model a problem through fuzzy logic principles, first of all the variables (cause-reasons-inputs and effect-results-outputs) must be attached with meaningful words such as rainfall, benefit, income, resistance, temperature, etc. For example, in a disarmament model finance, armament, enmity, weapon quantity and modernity of the weapons etc. are the variables. Each one of these words includes uncertainty in terms of vagueness, incompleteness, doubt, and incompleteness. They can be sub-classified in model construction.

The second question is which variables are inputs or outputs? This is tantamount to identify antecedent and consequent variables. It is similar to any mathematical expression, where there are dependent and independent variables. In the disarmament problem "armament" is the consequent variable and other four are antecedent variables. In the classical logic, each variable exists due to its opposite like armament-disarmament, support-not support, modern-not modern and alike. Such two alternatives are demolished in the fuzzy logic and instead more general classifications with middle classes are taken into consideration as multiple alternatives. Thus there are many relationships between sub-classes.

Sub-classifications should be completed for each variable. They can be sub-classified according to the following.

(1) "armament" ("little", "middle", "enough", "much", "very much") as the consequent variable;
(2) "finance" ("little", "middle", "much") as one of the antecedent variables;
(3) "enmity" ("small", "middle", "big", "very big") as another antecedent variable;
(4) "weapon quantity" ("little", "middle", "much") as another antecedent variable;
(5) "modernity" ("classic", "middle", "modern") as one of the antecedent variables.

The third stage in fuzzy modeling seeks answer to the question how many propositions can be generated from the classification of the antecedent variables? In the example given above, there are four antecedent variables each with in sequence 3, 4, 3 and 3 sub-classes, which make the number of proposition alternatives as = 108. This is a mechanical way of writing down the proposition numbers even though some of them are not valid.

As stated earlier in this chapter, in the fuzzy logic propositions antecedent variable sub-classes are brought together with the use of "AND" conjunctive. Additionally, different propositions are brought together in the form of wholeness by "OR" logical conjunctive. The collection of these propositions is referred to as the "rule base" of the problem at hand. Let us now ask whether an engineer should run after rule base or data base for the construction of suitable model. Those attached with classical and symbolic logic outputs will definitely run after data base without caring for the rule base, after all they think that the rule base is imbedded in the formulations, and there is no need for their revision. If revision is necessary, engineer might not be trained for such a task thought philosophy of engineering and critical assessment of existing formulations. Unfortunately, in classical engineering training nobody cares for rule base. However, data base is always at the top of modeling agenda. Some of the fuzzy logic rule statements can be written down as follows concerning the problem of disarmament.

Rule 1: *Finance* is "middle" AND *enmity* is "big" AND *quantity* "little" AND *modernity* is "classical" THEN…
OR
Rule 2: *Finance* is "" AND *enmity* is "middle" AND *quantity* "little" AND *modernity* is "classical" THEN…
OR
Rule 3: *Finance* is "middle" AND *enmity* is "big" AND *quantity* "little" AND *modernity* is "classical" THEN…
OR
"   " " " " " " " " " " " THEN…
"   " " " " " " " " " " " THEN…
OR
Rule 108: *Finance* is "much" AND *enmity* is "very big" AND *quantity* "much" AND *modernity* is "classical" THEN…

Fig. 4.4 Thinking gradients

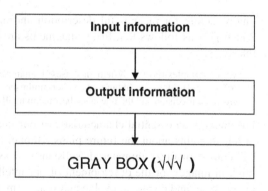

Two important points in these propositions are that the antecedent variables in italics are related to each other in a sequence by "ANDing" and adjectives in quotation marks imply uncertainty in a verbal form. Each one of these propositions (rules) expresses a piece of information from the wholeness of the phenomenal explanation. Each rule in a way divides four dimensional verbal input domains into sub-domains again verbally. Up to now, there is no numerical value involved in writing down the rule set.

Different rules are combined together for explaining the wholeness of the phenomenon concerned by "ORing". In short, the fuzzy rules have antecedents that are connected by "ANDing" and rules are connected by "ORing". One should notice that "AND" and "OR" logical conjunctives do not change with logic system.

4.10.1 Fuzzy Logic Thinking Stages

Figure 4.4 indicates the steps necessary for the completion of thinking process. Each step cannot be explained in a crisp manner and each individual depending on his/her capabilities may benefit from this sequence.

Although human wonder and mind are the sources of fuzziness, they also serve overcoming problems with human experience, which can be regarded as expert views. The fuzzy concepts in understanding complex problems are dependent on observations, experiences and conscious expert views. When problems are solved, there is always fuzziness attached to them, which paves ways for future developments. Hence, the scientific solutions cannot be taken as absolute truths in positivistic manner.

Conscious direction of attention towards an external object causes the object to be perceived by mind into the realm of fuzziness, which gives rise to a chain of perception, experience, feeling, thinking, understanding, knowing, and finally, acting for meaningful description and analytical solution. It is stated by Zadeh [13] as follows that fuzzy statements are the only bearers of meaning and relevance.

> As the complexity of a system increases, human ability to make precise and relevant (meaningful) statements about its behavior diminishes until a threshold is reached beyond which the precision and the relevance become mutually exclusive characteristics.

Fuzziness is an essential characteristic of our imaginations that raise and dissolve in our thoughts about the future plans. Human thoughts have blurred boundaries and consist of fuzzy immaterial 'substance'. Having in mind how important is to think in images for the development of our intelligence and capacity to learn and know, to act and create, to evolve and transform, one should not underestimate the role of Fuzziology [2].

At this stage it is useful to mention about the three stages of human thinking in the Middle Eastern philosophy for reaching to a solution of any problem in general (Chap. 2). These three words are "takhayyul" (imagination), "tasaw-wur" (geometric configuration) and "tafakkur" (idea generation). Any external object whether it exists materialistically or not, human beings try to imagine its different properties in a fuzzy world. This gives him/her the power of initializing individual and personal thinking domain with whole freedom in any direction. After the object becomes into existence vaguely in the mind, then it is necessary to know its shape, which is related to geometry. It is essential that the geometric configuration of the phenomenon must be visualized in mind in some way even though it may be a simplification under a set of assumptions. Again the fuzzy shapes in the mind are put down as crisp geometrical shapes such as square, triangle, circle, ellipsoid, etc. or their mixtures for the mathematical treatments.

In 1932 Gödel [3] proved that in any axiomatic mathematical system (theory), there are fuzzy propositions, that is, propositions which cannot be proved or disproved within the axioms of this system.

At the moment when one consciously directs his/her attention towards an external object, the object enters into the realm of his/her fuzziness of perception, which includes feeling, thinking, and understanding, experiencing, knowing and applying. The levels of fuzziness correspond to the levels of his/her capacity for understanding depending on the levels of consciousness.

4.10.2 Imagination and Perceptions

A fuzzy perception is an assessment of a physical condition that is not measured with precision, but is assigned an intuitive value. No matter how good the measuring equipment is, everything in the universe has fuzziness to a certain extent. The fuzzy perceptions can serve as a basis for processing and analysis in a fuzzy logic control system.

A fuzzy set is a group of anything that cannot be precisely defined. Consider the fuzzy set of 'discharges.' How big is a discharge? Where is the dividing value between "low" and "medium" discharge? Is a 50 m^3/s discharge a "low" or "moderate" discharge quantity? How about 75 m^3/s?, What about 156 m^3/s? The assessment should take into consideration the expert view. Other examples of fuzzy sets are "hazardous flood", "short duration drought", "warm days", "high hydraulic head", "small drainage area", "medium viscosity", "brackish water", etc. For an analysis, it is necessary to have a way how to assign some rational value to intuitive assessments for individual elements of a fuzzy set. The human fuzziness must be translated to numbers that can be used by a computer. This can be done by the assessment of a value from 0 to 1. For the question of

how severe is the flood?

the human might rate it at 0.3, if the flood danger is low. The expert might rate the severe flood at 0.9, or even with 1.0 membership degree, if it is in winter season. These perceptions are fuzzy, just intuitive assessments, not precisely measured facts.

By making fuzzy evaluations, with zero at the bottom of the scale and 1.0 at the top, one can have an analysis rule basis for the fuzzy logic method, and s/he can accomplish analysis or control project. The results seem to turn out well for complex systems or systems where human experience is the only base from which to proceed, certainly better than doing nothing at all, which is where one would be if unwilling to proceed with fuzzy rules.

Human common sense is either applied from what seems reasonable for a new system or from experience for a system that has previously had a human operator. Here is an example of converting human experience for use in an engineering system. Water engineers are not able to automate with conventional logic. Eventually, they translate the human "perception" into lots of fuzzy logic "IF-THEN" rules based on human experience. Reasonable success was thereby obtained in automating the plant. Objects of fuzzy logic analysis and control may include physical control, such as flow speed, or operating a dam, financial and economic decisions, psychological conditions, physiological conditions, safety conditions, security conditions, and much more [10].

Human beings have the ability to take in and evaluate all sorts of information from the physical world they are in contact with and to analyze mentally, average and to summarize all these input data into an optimum course of action. All living things can do this, but humans do it more and in a better manner.

If much of the information is not very defined precisely then it is called fuzzy input. However, some of the input might be precise and non-fuzzy such as the raingages' readings in definite numbers. The processing of all this information is not very precisely definable, and therefore, it is called as fuzzy processing. Fuzzy logic theorists would call it fuzzy algorithm usage (algorithm is another word for procedure or program as in a computer program).

4.10.3 Fuzzy Reasoning

Reasoning is the most important human brain operation that leads to creative ideas, methodologies, algorithms and conclusions in addition to a continuous process of research and development. Reasoning stage can be reached provided that there is stimulus for the initial triggering of mental forces. Triggering of pondering on a phenomenon comes with the physical or mental effects that control an event of concern. These effects trust imaginations about the event and initial geometrical sketches of the imaginations by simple geometries or pieces and connections between them [10]. In this manner, the ideas begin to crystallize and they are conveyed linguistically by means of a native language to other individuals to get their criticisms, comments, suggestions or support for the betterment of the mental thinking and scientific achievements.

Approximate reasoning helps to resurface in information technology, where it provides decision support and expert systems with powerful reasoning bounds by a minimum of rules and it is the most obvious implementation for the fuzzy logic in the field of artificial intelligence. It is already explained how one can easily relate logic to ambiguous linguistics in forms of different fuzzy words such as "very", "small", "high", and so on. Such flexibility allows for rapid advancements and easier implementation of projects in the field of natural language recognition. Fuzzy logic brings not only logic closer to natural language, but closer to human or natural reasoning. Many times knowledge engineers have to deal with rather vague and common sense descriptions of the reasoning leading to a desired solution. The power of approximate reasoning is to perform reasonable and meaningful operations on concepts that cannot be easily codified

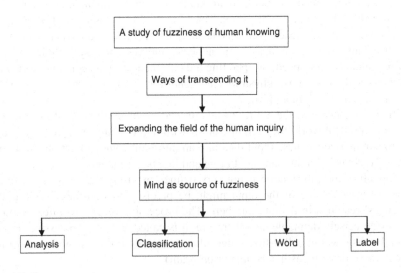

Fig. 4.5 Fuzzy concepts

using a classical approach. Implementing the fuzzy logic will not only make the knowledge systems more user friendly, but it also will allow programs to justify better results.

One of the powerful fuzziness sources in human ever-emerging desires of various kinds originates from simple physical desires, which are shared with other creatures to much more complicated desires that are specific for human nature. Every desire agitates the mind and distracts the process of concentration indispensable for an act of understanding to be productive. The stronger an emergent desire, the higher is the degree of agitation it stirs up, the less is the degree of concentration of mind and the less is the degree of concentration, the fuzzier is the process of thinking, the lower is the degree of understanding. Most of the desires self-propel their intensity (the more one tries to satisfy them, the higher becomes the demand and the way of moderation) the "middle way" is hard to follow, when the fire of desires is burning inside thinker and making the minds restless, turbulent and obstinate. Amidst of such feelings, the human mind is completely free in thinking including every extreme towards any direction. This is referred to as the philosophical thinking, which must be filtered later through the logic rules for deducing proper, meaningful and useful statements (propositions, premises) leading to plausible conclusions.

The restlessness and turbulence of minds are permanently intensified by the stress in which one lives due to the competitiveness (rivalry) inherent in today's society and the helplessness of majority to get out of the social boxes and cages (in which one has been pushed by economic forces too strong to be withstand), even if one desperately desires to. Although the strength of passion, with which one pursues truth and understanding, is a powerful stimulator and "energizer" of thinking, understanding needs "peace of mind", a mind, which is calm and cool, composed and collected. Paradoxically enough, while being sources of fuzziness, mind and desires are, at the same time, key factors for overcoming (transcending) it, especially if it relates to problems deeply rooted in human experience.

The fuzzy concepts in problem understandings that emerge out of life complexity as it unfolds cannot be resolved at the same level of knowledge that one has when these problems appear. Only when our consciousness is expanded to a higher level than the tension fades away and the problems, being seen in a new light, are no longer problems. When problems dissolve, one can say that the fuzziness related to them has been transcended.

Every isolated phenomena and processes in engineering have fuzzy knowledge content always. The significance of fuzziness opens ways for changes, evolution, growth, and continuous scientific developments. Figure 4.5 gives the document of steps in fuzzy thinking and problem solving [9].

The qualitative jump of consciousness to a higher level results in transcending the fuzziness. As far as consciousness is of a holistic characteristic of human, and perhaps not just of human, but also nature and not only a product of mind, but its growth and transformation are possible when the factors responsible for the integrity of all three inseparable constituents of human individuality, which are body, mind and soul, become simultaneously activated. This simultaneous activation ("firing" or "triggering") is referred to as a consciousness resonance and hence,

The fuzziness of understanding can be transcended when the consciousness resonance occurs.

The consciousness resonance is a resonance of all factors responsible for human integrity as manifested in the holistic nature of consciousness. What are these factors? First, factors, which contribute in keeping human body healthy and human mind capable to think and decide, no matter what kind of logic it prefers—fuzzy, binary, inductive, deductive, abdicative, etc. However, these factors are not enough. The consciousness resonance cannot occur when neglecting the soul factors; one can name some of them as sensitivity and responsiveness, awareness and ability to stay awake, passionate desire to get out of the "attractor" of egocentric thoughts and desires, compassion and love, willingness to explore more subtle and spiritual dimensions of reality and to share with others skill, knowledge and wisdom.

The consciousness resonance does not eliminate fuzziness, which is an eternal companion to any process of thinking and knowing. At the same time, when the consciousness resonance helps us to transcend the fuzziness related to a problem that dissolves, it opens space for new problems to emerge bringing with them new fuzziness to puzzle our thoughts and feelings. At any level of consciousness, there are infinite number of phenomena and processes challenging the "swarm" of our perceptions, of our beliefs and hopes, views and attitudes, aspirations and dreams.

The term resonance has a clear meaning in physics—it is a process of initiating a vibratory response in a receiver that is attuned to an emitter, which is considered as a source of vibrations (causal facts) they can be periodic, aperiodic or chaotic. In the process of resonance, these vibrations "fire" sympathetic vibrations in the receiver (consequent), the magnitude of which is often greater than the magnitude of the vibrations generated by the emitter.

The reasoning (philosophy of fuzzy thinking) is based on graded concepts. It is a concept in which everything is a matter of degree, i.e. everything has softness (elasticity). The fuzzy logic theory has been given first in its present form through the early publication by Zadeh [12]. He wanted to generalize the traditional notion of a set and a statement to allow the grades of memberships and truth values, respectively. These efforts are attributed to the complications that arise during physical modeling of real world. These are,

- Real situations are not crisp and resolute; hence they cannot be described precisely,
- The complete description of a real system often would require by far more detailed data than a human being could ever recognize simultaneously, process and comprehend.

Zadeh calls the last statement as the "principle of incompatibility". Its message is that the closer one looks at a real-world problem, the fuzzier becomes its solution.

Finally, all the conclusions must be expressed in a language, which can then be converted into universally used symbolic logic based on the principles of mathematics, statistics or probability statements. This explanation shows that fuzzy logic is followed by symbolic logic (mathematics). Unfortunately, in many educational systems all over the world, this sequence of language and symbolism is overturned into the sequence of first symbolism (mathematics) and then partial linguistic understanding which is against the natural reasoning abilities of human. This is especially true for countries or societies who are trained with symbols and those when they return to their community, the first difficulty is to convey the scientific messages in his/her language, and therefore, in order to avoid such a dilemma the teacher bases the explanation on symbolic logic. This is one of the main reasons why scientific thinking and reasoning are missing in many institutions all over the world. The avoidance of such a problem is approximate reasoning where the facts are explained through natural languages first (Chap. 2).

The subjectivity, i.e., dependence on personal thoughts is the greatest at the perception stage and as one enters the visualization domain, the subjectivities decrease and at the final stage since the ideas are exposed to other individuals, the objectivity becomes at least logical, but still there remains some uncertainty (vagueness, incompleteness, missing information, etc.), and hence, the final conclusion is not crisp but fuzzy. Fuzzy reasoning in many engineering always exists, but in the classical and mechanical approaches they are deleted artificially by idealizations, isolations, simplifications and assumptions.

The classical logic renders the final stage in solutions into crisp forms by defuzzification, which means neglecting all the uncertainties either through the assumptions or through a safety factor or confidence interval in many engineering solutions (Chap. 2). Crisp reasoning conclusions do not provide soft domain for further research especially in many engineering aspects that are involved with nature. Therefore, classical methodologies and formulations are fragile, hard and difficult to accept the consequences. In order to avoid the crispness, the statistics and axiomatic probability concepts are suggested, but they are also based on the classical logic, where the consequences are black and white without gray tones, which is available in approximate reasoning through fuzzy logic principles and modeling.

4.11 Logic and Rules

The meaning of logic has been explained already in the early sections and chapters; it implies thought or reason (logic) is the study of arguments, which express inferences, the processes whereby new assertions are produced from already established ones. Hence, a particular concern in logic is to form the structure of arguments in the forms of propositions. In logic structures epistemology plays a significant role for providing a rational mechanism towards the extension of presently available knowledge sources. Logic has been in very close touch with philosophy, but today it constitutes the bases of mathematics, and, even more recently, computer science. In its historical evolution, logic has gone through the stages of absolute causality and probability (crisp logic), mathematics (symbolic logic) and fuzziness (fuzzy logic).

Logic searches for the meaningful propositions among many philosophical sentences in any content. It is well appreciated that not all the philosophical sentences have logical structure and only logical sentences lead to thinking, possible interrelationships between various categories (cause-reason-input) and deduction (effect-result-output). It is, therefore, necessary to have some guidelines for the identification of logical statements in a given text or to construct them in the thinking process about the internal generation mechanism of the phenomenon concerned. The simple way of searching for a logical statement is to find one or more of the following logical words. These are,

1. "AND", this is the logical connective word which joints two categories or statements in such a way that they both are included in the final decision. This is also referred to as the "ANDing" process.
2. "OR", is another logical conjunctive that takes into consideration two categories but leads to a common deduction (decision) such that common parts of these two categories are the constituents of the deduction. It is also known in this book as "ORing" process.
3. "NOT", is the negation of whatever is the original classification. It is also described as "NOTing" process or operation in this book.
4. "IF...(A)...THEN...(C)...", is the argument that includes in the antecedent (A) part interrelationships among the causative variables leading to the consequent (C) part, which constitutes the final decision. Any statement that has the form of IF...THEN... is a part from the rule base.

A good statement is one whose conclusion follows from its antecedent part. The final decision as conclusion constitutes transfer through an inference system from the premises.

Classical logicians argue that fuzzy logic is unnecessary. Anything that fuzzy logic is used for can be easily explained using classic logic. For example, true (white) and false (black) are discrete. Fuzzy logic claims that there can be a gray degradation between true and false. Classic logic says that the definition of terms is inaccurate, as opposed to the actual truth of the statement.

4.12 Why Use Fuzzy Logic

As an esteem of the eastern thinking, philosophical objects may be raised by logical premises and implications along three basic mental activities, namely, imagination, conceptualization and deep thinking which jointly lead to idea generations [9]. Since the existence of terrestrial life, human beings have interaction with nature, which has provided the basic material in the form of objects and events evolving with time and space for the human perception and mental activity. At the early stages of human history or during the childhood of any individual, these stages play roles in different proportions and with experience they take final information generation forms. Each of these activities includes uncertainty because imagination, conceptualization and thinking stages are rather subjective depending on the individual's background. At any stage of human thinking evolution, the premises include to a certain extent uncertainty elements such as vagueness, ambiguousness, possibility, probability and fuzziness.

At the very elementary stages of mental thinking, activity objects are thought as members or non-members of a given or physically plausible domain of variability. This brings into consideration sets, which include possible outcomes or basis of the investigated phenomenon. In formal sciences such as physics, geology, etc., almost invariably and automatically, these elements are considered as either complete member of the set or completely outside the same set. Hence, the classical logic of pairs in the form of one or zero; positive or negative; yes or no, black or white, etc., are employed at the foundation of any scientific phenomena for mathematical modeling. However, Zadeh [12] suggested, instead of crisp membership consideration, rather continuity of membership degrees between 0 and 1, inclusively. Hence, fuzzy sets play intuitively plausible philosophical basis at every stage of the aforementioned mental activity chain (see Chaps. 3 and 5).

The guiding principle of fuzzy computing is to exploit the tolerance for imprecision, uncertainty, and partial truth to achieve tractability, robustness, and low solution cost. What makes fuzzy logic so important is the fact that most of human reasoning and concept formation are linked to the use of fuzzy rules. By providing a systematic framework for computing with fuzzy rules, fuzzy logic greatly amplifies the power of human reasoning [14].

(1) Fuzzy logic uses information efficiently; all available evidence is used and propagated until final defuzzification is robust to uncertain, missing or corrupted data.
(2) Fuzzy logic encodes human expert knowledge/heuristics, common sense, and the constraints are naturally enforced.
(3) Fuzzy logic systems are cheap, training data are not required, models or joint/conditional probability distributions are not needed.
(4) Relatively straightforward to design and implement.
(5) There is nothing fuzzy about fuzzy logic.
(6) Fuzzy logic is different from probability concepts,
(7) Designing the fuzzy sets is comparatively easier than any other sort of modeling.
(8) Fuzzy logic is stable, easily tuned, and can be conventionally validated.
(9) Fuzzy logic is a representation and reasoning process.
(10) Fuzzy logic is conceptually easy to understand.
(11) The mathematical concepts behind fuzzy reasoning are very simple.
(12) What makes fuzzy nice is the 'naturalness' of its approach and not its far-reaching complexity.

(13) Fuzzy logic is flexible.
(14) With any given system, it is easy to message it or layer more functionality on top of it without starting again from scratch.
(15) Fuzzy logic is tolerant of imprecise data. Every hydrologic event is imprecise if one looks closely enough, but most things are imprecise even on careful inspection.
(16) Fuzzy logic can model non-linear functions of arbitrary complexity.
(17) Fuzzy logic can be built on top of the experience of experts without data.
(18) In direct contrast to classical modeling styles, which take training data and generate opaque, impenetrable models, fuzzy logic lets one rely on the experience of people who already understand the system (expert views).
(19) Fuzzy logic can be blended with conventional modeling techniques.
(20) Fuzzy systems do not necessarily replace conventional methods. In many cases, fuzzy systems augment them and simplify their implementation.
(21) The basis for fuzzy logic is the basis for human communication, which is based on natural language, which is used by ordinary people on a daily basis and it has been shaped by thousands of years of human history to be convenient and efficient.

Generally, the fuzzy logic is recommended for the implementation of very complex highly non-linear processes, where a simple mathematical model cannot be obtained. It is not recommended for implementation to systems where simple, linear and adequate mathematical models already exist or where the conventional modeling theories yield a satisfying result. Fuzzy logic seems to be a general case for the classical logic and as such it does not present any better solutions for problems that might be easily solved using the "crisp" sets.

In recent years, interest in fuzzy logic approach in engineering modeling has increased but still there is not enough cover of all engineering aspects from fuzzy logic point of view. However, the number and variety of such applications are increasing, ranging from single engineering element to complex engineering system modeling. Natural phenomena have different types and varieties of internal and external factors that give rise to the occurrence of engineering phenomena in a sophisticated manner and such complications cannot be explained by classical crisp mathematical formulations. It is, therefore, very efficient to consider the basic philosophical fundamentals and logical foundations of these phenomena in order to reach reasonable conclusions. Although probabilistic, statistical and stochastic approaches are used to model the hydrologic processes, but they all depend on a set of assumptions and besides need data for the model establishment and verification. On the contrary, in fuzzy logic modeling neither a set of assumptions not the initial data availability is necessary for the model identification.

Fuzzy logic has its ingredients similar to the brain that deals with inexact information after the perception and fundamental inspirations about the phenomenon concerned. Fuzzy logicians have the opportunity for dynamic and numerical model free estimators for any engineering event. They are structured numerical estimators. Since, in engineering, estimations and predictions are the most required aspects, the fuzzy logic and system modeling suit engineering research and application easily. Fuzzy logic starts from highly formalized insights about the event structure and categorization that prevail in natural real world. Uncertain and imprecise information environments are the basis of the fuzzy thinking and logical deductions based on a subset of variables and their rational relationships. In this manner, complex and non-linear systems can be modeled without the involvement of extensive and rather difficult mathematical abstractions. The basic logical ingredients are translated into fuzzy IF-THEN rules as a kind of expert knowledge (Chap. 5). In this manner, any engineer can exploit his/her experience and expertise by lying down logical statements, which are referred to as the implications. In general, a fuzzy system combines the fuzzy sets and subsets of engineering variables with logical fuzzy rules so as to produce overall complex non-linear behavior.

Logical assessments require initial mental reasoning for the identification of similarities and differences between various scrap information, and finally, they are established in a systematic way to express the study conditions. Fuzzy logic approach leads to automation and machine intelligence of hydrologic phenomena.

The fuzzy logic can express all the subjectivity in human thinking and natural linguistic expressions in a comparatively undistorted manner. The following points lead to the necessity of fuzzy logic concepts use in the modeling.

- If at least one of the variables is continuous and cannot be digitized (break down into discrete segments) then the fuzzy logic approach must be used in modeling;
- If it is not possible to establish the mathematical model for the engineering event concerned;
- If the mathematical model takes a complex nature including multitude of equations;
- If the solution of the mathematical model is too complex for fast enough evaluation in real-time operations;
- If there is large memory requirements;
- If there are high ambient noise component in the evolution of the hydrologic event;
- If the process involves human interactions such as the opening of gates if the operator feels from his/her previous experience that the snow-melt will take place rather suddenly;
- If the expert can specify the set of rules for the operation of engineering system with rather vague information.

It is possible to execute the following hydrological processes with the use of fuzzy logic rule and system concepts.

- Control problem which is one of the most used problem solving in engineering;
- Time series components identification;
- Material resources operation and management;
- Database establishment;
- Estimation and prediction.

Human processing of information is not based on two-valued, off-on, either-or logic. It is based on fuzzy perceptions, fuzzy truths, fuzzy inferences, etc., all resulting in an averaged, summarized, normalized output, which is given by the human a precise number or decision value which he/she verbalizes, writes down or acts on. It is the goal of fuzzy logic control systems to also do this.

4.12.1 Fuzzy Logic

In order to cope with complex situations the fuzzy principles and modeling are justified, because such a modeling digests linguistic information in addition to numerical data. In many situations, linguistic information is obtainable through observations easier than numerical information. In any engineering study, there may not be numerical data, but engineer's observations provide a set of linguistic information that leads to logical and rational thinking with preliminary approximate deductions and solution rules. For instance, water taste by tongue gives expert information about the quality or looking at the rock outcrop provides the first impression about the infiltration rate and groundwater recharge. The historical traces of flood water level on both sides of a cross-section provide qualitative information about the past flood discharges. In solving real-life problems, engineering should not use only objective information (equations, algorithms and formulations) or only subjective knowledge (linguistic information) but s/he should exploit both information sources in arriving at an optimum solution. Fuzzy logic principles are extremely suitable for combining subjective knowledge with objective information.

Natural events are complex and arise from uncertainty in the forms of ambiguity. Subconscious human thinking digests complexity and ambiguity with partial solutions in natural, earth and social events. Although the complete description of a real phenomenon often requires detailed data, human perceptions and reasoning economize this requirement by pondering on the generating mechanism of the phenomenon concerned. This is due to the fact that human has the capability of approximate reasoning about the behavior of the phenomenon, which leads to generic understanding of the problem. As Zadeh (1973) quotes

> Complexity and ambiguity are related, the closer one looks at a real world problem, the fuzzier becomes its solution.

Oxford English dictionary defines the word fuzzy as 'blurred, indistinct, imprecisely defined, confused, vague', which gives the impression that there is no use of this word in daily life except with dangers and outside of the science and technology domains. However, the real situation is very opposite to such thinking and the fuzzy logic and system design became one of the most advanced methodologies of today. Fuzzy systems take the linguistic and verbal information as data and provide the answer accordingly in a vague manner, which includes the crisp solutions. This brings to the mind that although the basis of the methodology is fuzzy, the results are precise. It is better to start with fuzzy principles and to arrive at precise conclusions than to start with precise mathematical principles and to conclude with one crisp result, which may never appear in the real life.

Learning through observation, measurement, experience and reasoning about the complex phenomenon reduces the complexity in understanding. Reduction in the complexity means useful insights into the behavior of the problem with more certain ideas. As the complexity becomes marginal with little uncertainty, mathematical formulations provide precise description of the generating mechanism. In such cases the probabilistic, statistical and stochastic modeling in sciences become applicable for the description and even control of the phenomenon. Most often in many branches of science mathematical models are based on a set of assumptions, idealizations and simplifications in order to reduce complexity in an artificial manner so as to establish at least preliminary approximations that can be expressible by formulations, which are necessary to make quantitative and deterministic conclusions. Mathematical formulations are necessary in finding numerical solutions for the problem at hand. It must be remembered that any mathematical model is based on a set of restrictive assumptions, which overlooks the uncertainty in the problem.

On the other hand, for most complex systems with few numerical data where only ambiguous and imprecise information is available, none of the deterministic formulations in sciences help to solve the problem. However, fuzzy reasoning provides a way to understand and then interpret system behavior with interpolations based on the available scarce numerical but rather abundant verbal (linguistic) data about the generating mechanism of the phenomenon with its inputs (causes, antecedents) and outputs (results, consequents). In fuzzy reasoning, the information about the input and output variables is combined with logical insights about the system. This provides an ability to describe in words through a set of rules, the mathematical abstraction of the real world. Fuzzy modeling is to match ambiguous input and output information through fuzzy rules. It requires reasoning with logical footprints that constitute the backbone of the behavioral abstraction of the problem with rational and partial conclusions. Due to complexity and ambiguity, human ability provides inference by reasoning the internal mechanism of the problem, which requires not only crisp conceptions and mathematical formulations, but more artistic scenarios of different alternatives. Detailed explanation about fuzzy modeling in engineering aspects is presented by Ross [7].

Classical systems work with crisp and organized numerical data on the basis of two-valued (Aristotelian) logic, which has only two mutually exclusive alternatives like wet (white, yes, one, positive, true, etc.) and dry (black, no, zero, negative, false, etc.). Engineering sciences have almost in every corner gray fore and back grounds with verbal information, which is full of ambiguous, vague, imprecise and random information sources. It is a big dilemma how to deal with gray information for arriving at decisive conclusions with crisp and deterministic principles. Fuzzy logic principles with linguistic premises and vague categorization provide a sound ground for digestion of such information. The preliminary step is fuzzy logic conceptualization of engineering phenomenon with uncertainties in its input, system and output variables. Such an approach helps not only to visualize the relationships between different variables, but furnishes a philosophical detail about the system mechanism of the hydrological phenomenon without mathematical formulation.

Fuzzy set theory provides a rich and meaningful addition to two-valued logic. The mathematics generated by these theories is consistent, and fuzzy logic seems as the generalization of classical crisp logic. The applications which may be generated from or adapted to fuzzy logic are wide-ranging, and provide the opportunity for modeling of conditions which are inherently imprecise despite the concerns of classical logicians. Many systems may be modeled, simulated, and even replicated with the help of fuzzy systems, not the least of which is human reasoning itself.

It is emphasized in this book that fuzzy logic approach can provide the structure and solution procedure of engineering systems prior to any crisp (deterministic) method such as mathematics, statistics, probability or stochastic processes. In this way, engineer is able to develop creative and analytical thinking capabilities with the support of other expert views. Since, the modern philosophy of science insists on the falsification of current scientific results [6], there are always rooms for ambiguity, vagueness, imprecision and fuzziness in any scientific research activity. The fuzzy logic will attribute degrees to even a scientific belief (degree of verification or falsification) that assume values between 0 and 1, inclusively. Verifiability of scientific knowledge or theories by logical positivist means on the classical grounds that the demarcation of science concerning a phenomenon is equal to 1 without giving room for falsification. The conflict between verifiability and falsifiability of scientific theories includes philosophical grounds that are fuzzy. Although many science philosophers tried to resolve this problem by bringing into the argument the probability and at times the possibility of the scientific knowledge demarcation and scientific development, unfortunately so far the "fuzzy philosophy of science" has not been introduced into the engineering hydrology literature [8, 11]. Hence, it is the purpose of this book to give an account of fuzzy logic in the processes of scientific knowledge demarcation and progression. Dogmatic nature of scientific knowledge or belief, in the science as if it is not doubtful, is the fruits of formal classical Aristotelian logic, whereas fuzzy logic holds the scientific arena vivid and fruitful for future plantations and knowledge generation. Innovative engineering systems should lean more towards the basic scientific philosophy of the problem solving with fuzzy logic principles. Contrary to classical (two-valued) logic, fuzzy logic may be thought basically similar to a multi-valued logic. However, it is not exactly so due to uncertain boundaries between the multi subsets. It allows intermediate uncertain values to be defined between two-valued conventional evaluations like dry/wet, high/low, intense/sparse, etc. Notions like 'rather dry', 'highly humid' or 'semi-arid' cannot be formulated crisply except through FSs and models.

This book presents systematic and comprehensive modeling of uncertainty, vagueness, or imprecision through fuzzy principles and procedures for problem solving in engineering. There are several chapters for introduction to fuzzy logic containing basic definitions of fuzzy sets, clustering and fuzzy rule systems. It describes methods for the assessment of relevant rules in modeling engineering systems; verification and redundancy issues; and investigates rule response outputs, definitions and premises.

4.13 Engineering Geometric Description

The description of geometry is more significant that the mathematical conceptions, because engineering mathematics is based on the geometry of the concerned phenomenon. For instance, electrons are assumed to revolve around the nucleus on concentric circle trajectories. In reality, these trajectories are not existent, but the atomic problems cannot be solved without geometry, and therefore, the simplest geometry is assumed for the atomic structure. The final stage, tafakkur (idea generation), corresponds to concluding useful information deduction based on logic, which is always fuzzy but for the modern scientific requirements it is defuzzified into a crisp result. Fuzziness is the processes of learning, generating hypotheses and proving theorems.

Almost all the preliminary scientific findings are related to geometry including art structures from the depth of the history and this indicates very clearly that preliminary human perceptions have attachments with geometry prior to any other subject such as mathematics. Any object explanation can be perceived through geometrical shapes rather than language solely. For example, even though "tree" is explained as having its roots in the ground with branching roots below the trunk, it is better to draw its shape on a piece of paper or on the ground for better explanation. As Plato, one of the earlier Greek philosopher, wrote at the front of his academia that "Who does not know geometry, cannot enter this academy". On the other hand, Muslim scholar Ibn Haldun in fourteenth century stated that "if a mind gets information through shapes (geometry) then it is not possible that it errs".

It is well known that geometry is related to engineering design. Most often objects take their names from geometrical shapes. The earth has shape of a ball, and therefore, most often it is named as sphere-earth, likewise atmo-sphere, lithosphere, hydro-sphere, etc. In any scientific approach, for solution, the first step is based on imagination with its shape as visualization in the mind and after some development; the first design is put down for further discussions and criticisms leading to a successive changes according to new improvements. Geometry provides a common basis in the form of design for visual inspection and development of ideas by successive improvements. For example, atomic model where the nucleus in the center is surrounded by electrons in concentric orbits is a very fundamental geometry, which has led to the generation of a set of ideas. In deed any artist, sculptor or engineer is successful as long as they are able to draw what they have in their mind into a shape, which provides triggering for visual and linguistic improvements. The geometry provides a means of knowledge transfer in the shortest possible manner to other individuals. Plans and projects in engineering are all among such geometrical productions, which helps to transform imagination and verbal information to concrete shapes. Even mathematics and physics develop on the basis of geometrical shapes. If an opportunity is provided then instead of mathematical courses, geometry basic information should be given especially to engineering students in terms of Euclid, descriptive, spherical and cylindrical, Riemann and recently fractal geometries. It is also possible to regard geometry as the basis of basic sciences such as mathematics, physics, chemistry and biology. If geometry is given properly in any education system then the rate of research will increase significantly, because geometrical shapes help for visual understanding in addition to mental grasp and linguistic explanations.

An effective example has been provided for the use of geometry in problem solving about 1,000 years ago. A question about how the roots of a second order (quadratic) equation are found cannot be easily answered let along by engineers, but even by many academic staff rationally. However, ready answer appears as a result of memorization and all that remains for engineer or academician is the substitution of the constants in the second order equation in readily available formula, which yields two roots. If anybody is asked about the derivation of the memorized formulation, almost nobody can answer and explain it scientifically. Additionally, even the question who has invented it first?, will not find an answer. The solution has been found first by a Muslim scientist, Al-Khwarizmi, whose name is Latinized as "algorithm" in the ninth century by use of algebraic methodology. Automatically, the two roots of the second order equation,

$$ax^2 + bx + c = 0$$

can be found from the following formulation, which is in the memory of specialists without any logical, philosophical or rational explanations.

$$x_{1,2} = \frac{-b \pm \sqrt{b^2 - 4ac}}{2a}$$

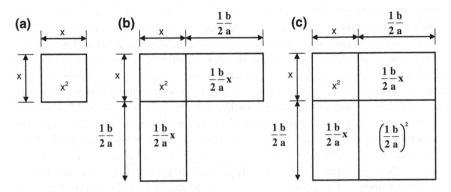

Fig. 4.6 Second order equation design

In the derivation of this equation the geometry is used as en engineering design procedure. This example provides a common basis for indications of how significant is the geometrical imagination, visualization and consequent design. Let us start with a general second order degree equation given as,

$$ax^2 + bx + c = 0$$

Division of both sides by a yields,

$$x^2 + \frac{b}{a}x + \frac{c}{a} = 0$$

A close inspection of the first term means linguistically that it is in the form of "square", and hence, a square with side length x is constructed as in Fig. 4.6a. Thus, x^2 represents the area of this square, which completes the geometrical interpretation of the first term. Now, let us try to give another geometrical meaning to the second term in this last expression. It has a rectangular shape with one side equal to x and the other to b/a. Herein, the question is which side of the square will be adapted as x and in order not to make distinction between the two sides, the rectangle will be divided into two halves with x side remaining the same. This thought leads to two rectangles as shown in Fig. 4.6b.

As a final step in Fig. 4.6, one can realize that it can be completed to another square with each side equal to (x + b/2a) as in Fig. 4.6c. The area of this big square is,

$$\left(x + \frac{1}{2}\frac{b}{a}\right)^2 = x^2 + \frac{b}{a}x + \left(\frac{1}{2}\frac{b}{a}\right)^2$$

On the right hand side, the first two terms is equal to $-c$ and consequently, one can write,

$$\left(x + \frac{1}{2}\frac{b}{a}\right)^2 = -\frac{c}{a} + \left(\frac{1}{2}\frac{b}{a}\right)^2$$

If the square root of both sides are taken one can obtain,

$$\left(x + \frac{1}{2}\frac{b}{a}\right) = \pm\sqrt{-\frac{c}{a} + \left(\frac{1}{2}\frac{b}{a}\right)^2}$$

After the necessary arrangements the final solution becomes as,

$$x_{1,2} = \frac{-b \pm \sqrt{b^2 - 4ac}}{2a}$$

This procedure shows that depending on the linguistic information and geometry, the solution of the second order equation solution is possible simply, rationally, logically and geometrically. After all these steps one can arrive at the mathematical formulation solution with symbols representing unknown, x, and constant parameters, a, b and c. Who knew such a simple and plausible solution from engineers or academicians? Perhaps 99 % of all specialists are not aware of this philosophical and logical solution, but the consequent equation is useable all over the world based on memorization without critical debate. This also shows the difference between non-generative (memorization) knowing and rationally (logically) knowing. One way is static, memorization, dogmatic, transferable and without mind function and the other alternative has dynamism, critical assessment, linguistic debate, rational and logical reasoning.

The author insists that all formulations in engineering have similar background based on linguistic rationality, philosophical thinking and logical principles with geometrical imagination and visualization. For the linguistic information and knowledge to be generative, philosophical thinking together with logical inferences play significant roles. If the linguistic dynamism is applied to known formulations and equations then one can eliminate all deficiencies, approximations, invalid segments through clearness, selectiveness and transparency to the memory. This is similar to debugging during any computer program development and clearing it from possible illogical statements or viruses. It must be remembered always that any formulation or equation derivation has gone through such rational, philosophical and logical processes, and still innovative products will also be subjected to similar processes for the advancement and improvement of engineering end products. However, engineering thoughts as mentioned before have to have limited boundaries for the practical applications at some time, which can be modified during some similar applications. Expansion of the boundaries is possible by adjustment by considering linguistic arena of knowledge and information.

4.14 Engineering Mathematical Description

In engineering there are many rational formulas that can be deduced by simple geometric imaginations and descriptions similar to what has been explained in the previous section. An engineer with rational thinking principles is capable to deduce simple but effective equations that can be employed in practical applications. Such simplifications should be at the service of engineers at any time and even without firm background simplifications and rational reasoning may lead to practically useable products.

Fig. 4.7 Wire resistance

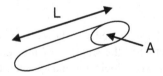

Let us think about how resistance of a wire gives way to information production. Rational inferences will be considered and at the final stage we will see that the linguistic words after their representation by symbols will lead to a formulation with a parameter of which the numerical value can be determined by experimentation only. An engineer should be aware of the fact that not only rational inferences but also their support by experimentation can lead to useful formulations. Let the resistance of a wire denoted by R, and then ask what the major variables are that affect the resistance? This question can be explored further by considering the geometrical dimensions of the wire, which are its cross sectional (circular, the simplest form) area, A, and length, L as in Fig. 4.7.

After rational thinking one can write implicitly the dependence of R onto L and A as R = f(A, L), which means that the resistance is a function of (i.e. depends on) the cross-sectional area and the length, which represent the geometric dimensions of the wire. If one reasons rationally about how these geometric quantities affect the resistance then s/he can realize that the relationship between R and A is inversely proportional, whereas R is directly proportional with L. This linguistic and logical relationship (in a way correlation) statements enables one to write the following combined proportionality.

$$R \, \alpha \, \frac{L}{A}$$

Herein α implies proportionality. Rational thinking does not indicate equality, but if one wants to write equality, then a constant must be imported on one of the sides, for instance, here the constant c is added on the right hand side and finally,

$$R = c \, \frac{L}{A}$$

This expression is derived with linguistic, philosophical thinking and rational reasoning without any numerical thoughts. Now the question is what are the meaning, interpretation and relevance of the constant? In order to answer this question, the constant can be left as subject as follows.

$$c = R \frac{A}{L}$$

Fig. 4.8 Area-rainfall relationships

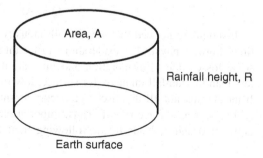

This expression tells us much verbal information that may not be thought at the first glance. Let us list some of these interpretations as follows.

- Provided that the area and length of the wire are assumed equal to unity, then one can say that the constant is equal to the wire resistance. This is to say that the constant is the resistance per unit area per length.
- In the case of unit area and length, the volume is also equal to unity, thus the constant is the resistance of unit wire volume. This is the case, which can be named as specific resistance that is equal to the constant.

After all the aforementioned geometrical and linguistic explanations in the formulation derivation, one can also reason this expression from physical point of view. Any wire with high resistance will cause increase in its temperature during the transmission of any signal. Increase in the temperature will lead to light emissions. Logically, after rational reasoning one can interpret that the smaller the cross-sectional area of the wire, the more will be its illumination and temperature. Also the longer the length, the more will be the illumination and temperature. These interpretations indicate the reason why in any lamp the spiral of wire has very small cross-sectional area.

Another rational inference that is frequently used in engineering is flood water volume per time (discharge) calculation. This approach is referred to as the rational formula. Again in the form of a black-box modeling, without caring what goes on internally, plausible relationship is sought between the input and output variables. If the flood discharge is denoted by Q then the implicit relationship is $Q = f(A, R)$, where A is the water collection area (catchment area, drainage basin are) and R is the rainfall height. Rational thinking may bring this expression in an explicit form by using philosophical principles along with logical inferences. First let us consider the geometry of the case as in Fig. 4.8.

According to Fig. 4.8 the volume, V, of rainfall over the whole area can be expressed as,

$$V = AR$$

Herein, an implicit assumption is that rainfall is uniformly distributed over the area and it is assumed to fall during a certain time interval, T, then division of both sides by T in the previous expression leads to new definitions as discharge $Q = V/t$ and rainfall intensity, $I = R/T$; and hence,

$$Q = AI$$

It should be noticed that the rainfall intensity is equal to the rainfall height per time. Further physical considerations bring into the view that there will appear some losses due to evaporation and infiltration. The total loss amount must be less than the rainfall intensity for surface flow to take place, and hence, discharge. It means that not all the discharge occurs as surface flow, but its certain fraction will appear as surface runoff. Thus, importation of a coefficient C with its value between 0 and 1, leads to the following final formulation that can be used for

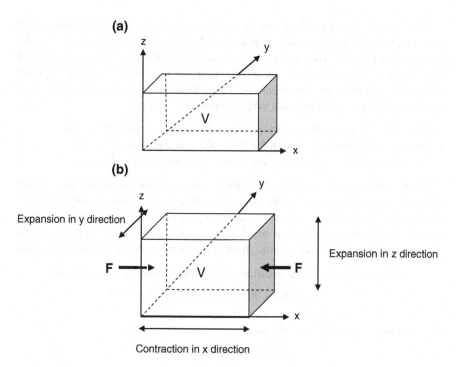

Fig. 4.9 Incompressible material instantaneous deformations

practical calculations provided that all the quantitative amounts on the right hand side are known.

$$Q = CAI$$

It is now the turn for interpretation of the constant. If it is made subject one can write,

$$C = \frac{Q}{AI}$$

which shows that C is a dimensionless factor. It is the amount of discharge per unit area per unit rainfall intensity. This coefficient is referred to as the runoff coefficient.

Another example can be given for derivation first order partial differential equation without mathematics, first linguistically and then translation of the linguistic sentences into mathematical notation. The same notation can be expanded to n dimensional space by considerations of physical meanings, unit homogeneity, initial and boundary conditions. Let us imagine that there is a parallelized prism volume, V, made of incompressible material as in Fig. 4.9a. Herein, "incompressibility" is the main assumption.

Let force, F, be applied as shown in Fig. 4.9b to the original volume as in Fig. 4.9a instantaneously, which implies that the time factor is not effective in the deformation process. The reader can think that the force is uniformly distributed over the two vertical faces of the prism. After the imagination, consideration of geometrical descriptions' comparison in Fig. 4.9a and b prior to and after the force application leads to the following verbal statements.

- There is a volume difference in the form of contraction along the x axis;
- There is a volume difference in the form of expansion along the y axis;
- There is a volume difference in the form of expansion along the z axis.

During the process since the material is assumed incompressible, the volume does not change, and hence, there is no difference in total volume, which implies notationally, $dV = 0$. Although the volume did not change, but the shape changed, and therefore, volume changes along each axes should add up to zero. It means that the summation of directional volume changes is zero or explicitly in linguistic terms,

(Volume difference in x direction) + (Volumedifference in y direction)

+ (Volume difference in z direction) = 0

which can be written notationally as,

$$dV_x + dV_y + dV_z = 0$$

However, one can write this last expression more explicitly as,

$$\frac{dV}{dx} + \frac{dV}{dy} + \frac{dV}{dz} = 0$$

this is the final expression for the problem. V is dependent on more than one independent variable (x, y and z), in other words V is partially dependent on x, y and z; in order to make this distinction mathematical consensus suggests using the following notation just to imply these partialities.

$$\frac{\partial V}{\partial x} + \frac{\partial V}{\partial y} + \frac{\partial V}{\partial z} = 0$$

This is a first order partial differential equation that is derived linguistically after a sequence of imagination, description (geometry) and idea generation without any mathematical involvement at initial stages. In this manner, even those who do not have a formal differential equation concept can grasp the basic process linguistically. In this manner, the gap between linguistically and mathematically oriented engineers has been bridged with no confusion. This example shows the fundamental significance of linguistic thoughts (philosophy) and subsequent logic. In open literature, this last expression is known as continuity (mass balance) equation, which is known by many engineers by memorization but they may not know the linguistic background garden in the derivation of it. If one does not know linguistic arguments then how s/he can debate with others. Is it necessary to get a formal education for the derivation of verbal equations or rational mind reasoning and logical inferences are sufficient to reach the final formulation?

The last expression is in three dimensions (3D) and now one can generalize it if there were four or more, say, n dimensions. Let us consider the fourth dimension time, t, in the sense that the force is not applied instantaneously but gradually. According to the aforementioned generalization one can write,

$$\frac{\partial V}{\partial x} + \frac{\partial V}{\partial y} + \frac{\partial V}{\partial z} + \frac{\partial V}{\partial t} = 0$$

this seems rational and logical inference. However, unit analysis is necessary at this stage by asking about the units of each term in this expression. The unit of volume is L^3, say for instance, m^3 and the unit of x, y and z directions is in L; unit of t is T. Accordingly, the first three terms in the expression will have unit of L^2, whereas the last term has L^3/T, which indicates that there is not unit homogeneity in the equation. In its present form it implies addition of apples and pears, which is not possible. In order to alleviate this situation either L^2's must be converted to L^3/T unit or vice versa. It is simple to convert L^3/T into L^2 through a conversion factor, α, and hence one can write,

$$\frac{\partial V}{\partial x} + \frac{\partial V}{\partial y} + \frac{\partial V}{\partial z} + \alpha \frac{\partial V}{\partial t} = 0$$

The unit of α turns out to be T/L which satisfies unit homogeneity. T/L is just the inverse of L/T, which implies in physics velocity, and therefore, α can be given a physical meaning as the inverse of velocity.

Another example for transition from linguistic statements to mathematical notations and relevant equation is the sentence, which states that:

the more money one has, the more s/he spends.

This statement implies a mathematical model that can be identified after the application of the philosophical thought steps in sequence. First, this sentence leads one to imagine about the spending phenomenon, where there is some money as quantitative material. Another scientific quantity is the word "spend", which implies change of that money with respect to time. One can describe the phenomenon as input variable money, M, and output variable as spending, dM/dt. The relationship between these two variables can be constructed by mental thought, because a slight pondering indicates that the spending is directly proportional with the amount of spending, which implies that

$$\frac{dM}{dt} \, \alpha \, M$$

where α is the proportionality sign. This expression can be converted into an equation form by importing a proportionality constant, say, c, then one can rewrite,

$$\frac{dM}{dt} = cM$$

This is the expression in linguistic terms that indicates the relationship of spending to money.

Above examples and explanations obviously show that engineering formulations can be derived with rational reasoning and logical principles. Thus, similar to the philosophy of science, philosophy of engineering fundamentals is important because they are the collections of critical reasoning, rational thinking, proposition suggestions and logical inferences.

The dire of "publish or perish" brought in front the scientific publications even in many top journals along a wrong path with mechanical flavor only. Although the rate of publication increased, the quality decreased, especially in areas where the most recent intelligent methodologies are used without logical care, but mechanical desires through software.

Early humans were thinking in an entirely uncertain environment for their daily and vital activities. It is possible to say that early knowledge and information were concepts derived from frequent observations and experience. Throughout the centuries, human thinking had support from scripts, drawings, calculations, logic and finally mathematical calculations. In the meantime, science is separated from philosophy with its own axioms, hypotheses, laws and final formulations especially after the renaissance in the seventeenth century. It is possible to state that with Newtonian classical physics, science entered almost entirely deterministic world where uncertainty was not even accounted among the scientific knowledge. However, today almost in all the branches of science, there are uncertainty ingredients and many scientific deterministic foundations become to take uncertainty form of fuzzy modifications. Among such conceptions are quantum physics, fractal geometry, chaos theory and fuzzy logical principles. However, some others such as the geological sciences have never gone through the stage of determinism, but unfortunately, in many institutions all over the world, deterministic educational systems affected the various training systems. With the advancement of numerical uncertainty techniques such as probability, statistics and stochastic principles scientific progresses in quantitative modeling had rapid developments, but still leaving aside the qualitative sources of knowledge and information, which can be tackled by the fuzzy principles only.

Recently, famous philosophers and scientists alike, started to spell out the uncertainty and fuzzy ingredients that are essential basis of scientific progress. For instance Russell stated that:

> All traditional logic habitually assumes that precise symbols are being employed. It is, therefore, not applicable to this terrestrial life but only to an imagined celestial existence.

As an esteem of the eastern thinking, philosophical objects may be raised by logical premises and implications along three basic mental activities, namely, imagination, conceptualization and subsequently idealization with jointly leading to idea generations. Since the existence of terrestrial life human beings have interaction with the nature which has provided the basic material in the form of objects and events evolving with time and space for the human mental activity chain. At the early stages of human history or during the childhood of any individual these stages play roles in different proportions and with experience, they take final forms. Each one of the chain element in the thinking process includes uncertainty because imagination, conceptualization and idealization stages are rather subjective depending on individuals. At any stage of human thinking evolution the premises include to a certain extent uncertainty elements such as vagueness, ambiguousness, possibilities, probabilities and fuzziness. Implication of mathematical structure from the mental thinking process might seem exact, but even today it is understood as a result of scientific development that at every stages of modeling, physical or mechanical, there are uncertainty pieces, if not in the macro scale, at least at the micro scale. It is clear today that mathematical conceptualization and idealization leading to satisfactory mathematical structure of any physical actuality is often an unrealistic requirement. As Einstein stated:

> So far as the law of mathematics refer to reality, they are not certain. And so far as they are certain, they do not refer to reality.

At the very elementary stages of mental thinking, activity objects are thought as members or non-members of a given or physically plausible domain of variability. This brings into consideration sets which include possible outcomes or basis of the investigated phenomenon. In formal sciences such as physics, geology, etc., almost invariably and automatically, these elements are considered as either completely members of the set or completely outside the same set. Hence, the Aristotelian logic of pairs in the form of one or zero; positive or negative; yes or no, black or white, etc., are employed at the foundation of any scientific phenomena for mathematical modeling.

References

1. Dimitrov V (1983) Group choice under fuzzy information. Fuzzy Sets Syst 9:25–39
2. Dimitrov V, Hodge B (2000) Why fuzzy logic needs the challenge of social complexity. In: Dimitrov V, Korotkich V (eds) With fuzzy logic in the new millennium. UWS, Richmond
3. Gödel K (1932) Ein Spezialfall des Entscheidungsproblem der theoretischen Logik. Ergebnisse eines mathematischen Kolloquiums 2:27–28. Reprinted and translated in (Gödel, 1986, 130–235)
4. Kilwardby R (1978) Commentaries on the Isagoge, Praedicamenta, Peri Hermeneias, Liber Sex Principiorum, and Liber divisionum. Lewry PO (ed) Robert Kilwardby's writtings on the Logica Vetus studied with regard to their teaching and method. PhD dissertation, University of Oxford, Oxford
5. Poinsot J (1955) The material logic of John of St. Thomas. University of Chicago Press, Chicago
6. Popper K (2002) The logic of scientific discovery (trans: Popper JFK, Freed L). Routledge Classics, New York
7. Ross T (1995) Fuzzy logic with engineering applications. McGraw-Hill, New York, pp 628
8. Şen Z (2002) Bilimsel Düşünce ve Mühendislikte Matematik Modelleme İlkel, Su Vakfı
9. Sen Z (2008) Bulanık Mantık İlkeleri ve Modelleme (Mühendislik ve Sosyal Bilimler) –Fuzzy Logic Principles and Modeling (Engineering and Social Sciences) Su Vakfı Yayınları,(Turkish Water Foundation Publication) pp 361

Chapter 5
Science and Engineering

5.1 General

As already explained in the previous chapters science is the human works towards finding, exploring, creating new ideas logically and symbolically about the behavior of objects about their spatio-temporal variations as materialistic quantities. At the end the behavioral cases are boiled down to algorithms, procedures and formulations in equation forms, which provide concentrated information by providing mutual relationships among the causative and responsive variables that are of interest for further researches or applications. Engineers are after the simple, economic, speedy and beneficial exploitation and application of the end products of scientific affairs. These definitions may give the impression that science and engineering may be rather mutually exclusive spheres of human activities, but they support each other especially since the last two decades. Engineering activities and technological developments need scientific principles and in the meantime scientific research and developments cannot be achieved without engineering plans, designs and structures. In general, although engineering education systems have scientifically oriented curriculum, but in practical training some surface information are given to engineering students without the internal scientific facts. Most often engineers are content with formulations, equations, algorithms and especially in the last 5–6 decades software started to dominate engineering affairs. Software includes all the logical and scientific steps in their body without scientific clues, but they execute the scientific and logically arranged steps speedily to provide numerical results for engineering design procedures.

Recently, expert views in parallel development with expert systems opened more scientific horizons for engineers, because expertness is by means of linguistic (verbal, language) statements, which can be judged in one's mind logically, and hence, scientific thoughts as triggering may appeal the engineer to be aware of some scientific realities about the problem.

In general, science is concerned with generation of scientific knowledge, which ripens in the academic circles including improvement of existing theories or suggestion

Z. Şen, *Philosophical, Logical and Scientific Perspectives in Engineering*,
Intelligent Systems Reference Library 143, DOI: 10.1007/978-3-319-01742-6_5,
© Springer International Publishing Switzerland 2014

of new ones, through scientific researches leading to papers, reports, M. Sc. and Ph. D. theses. Technological achievements include inventions leading to patents, know-how information, and blueprints. On the other hand, engineering achievements are concerned with material products and their strengths' consideration in many engineering structures. Technological community includes inventors, engineers, technicians, workers, managers, and common people with invention ambitions.

5.2 Scientific Sense and Thinking

Various phenomena in engineering, medicine and sciences take place in a complex world, where generally complexity arises from uncertainty in the forms of ambiguity. Scientists address problems of complexity and ambiguity at times subconsciously since they could think; these ubiquitous features pervade most natural, technical, and economical problems faced by the human race. The only way for computers to deal with complex and ambiguous issues is through rational and logical thinking, systemizing, controlling and selecting the most suitable solution among various alternatives as a result of decision making procedures.

Common sense dictates that some form of empiricism is essential to make sense of the world. In traditional quantitative educational training, the classical dualism as the unknown relationship between subjectivity and objectivity is often addressed by adopting an objectivist, empiricist or positivistic approach, and then by applying a scientific research design. Even based on crisp logic, scientific thinking starts in an entirely subjective medium. Subjective thinking penetrates objectivity domain by time through imagination and visualization, and hence, there is not a definite line between subjectivity and objectivity. Empirical works, which are based on either observations or measurements as experimental information, help to decrease the degree of subjectivity at the benefit of objectivity degree. In a way, none of the scientific formulations obtained up to now is completely crisp, but they are regarded as deterministic and crisp information provided that fundamental assumptions such as mutually exclusiveness and exhaustiveness are taken into consideration. Any deterministic scientific information can be relieved by modifying one of the basic assumptions. This implies that all the scientific principles are not completely deterministic, but include vagueness, incompleteness and uncertainty even to a small extend, and hence, they can be considered as either probabilistic numerically or fuzzy linguistically by nature or by human understanding.

As already mentioned in Chap. 2, any scientific thinking has three major steps, namely, imagination, visualization and idea generation. Imagination part includes the setting up of suitable hypothesis for the problem at hand and subsequent visualization stage is to defend the representative hypothesis. Scientists typically use variety of representations, including different kinds of figures (geometry) to represent and defend the hypotheses. Scientific hypothesis justification is possible only through the understanding of visual representation, and if necessary, modification of the hypothesis should be in progress. On the basis of hypotheses, the scientists behave as a philosopher by generating relevant ideas and their subsequent dissemination, which should include new and even controversial ideas, so that other scientists can overtake and elaborate more on the basic hypotheses. Whatever are the means of thinking, the scientific arguments are expressed by verbal expressions prior to any symbolic and mathematical abstractions. Especially, in engineering and physical sciences visualization stage is represented by algorithms, graphs, diagrams, charts and figures, which include tremendous amount of condensed verbal information.

The scientific visualizations are conducted with geometry since the very early beginning of scientific thoughts. This is the reason why the geometry was developed and recognized by early philosophers and scientists than any other scientific tools such as algebra, trigonometry, and mathematical symbolism. Al-Khawarizmi (died 840 A.D.) who is known in the west as his Latinized name "algorithm" solved second order equations by considering geometric shapes. For instance, he visualized x^2 as a square with side equal to x, and terms such as ax are considered as a rectangle with base length x and height equal to a. This geometrical thinking and visualization made him the father of "algebra". All his discussions were explained linguistically in Chap. 4.

All the conceptual models deal with parts of something that is perceived by human mind. Of course, among the meaningful fragments of the phenomenon there exist clear and hidden interrelationships, which are there for the exploration of human intellectual mind. Such possible relationships can be explained by a set of fuzzy rule statements as mentioned earlier in different chapters. Figure 5.1 indicates the fragments of thinking, sensations, thoughts and perceptions, which serve collectively to provide partial and distorted conceptual models of reality in representing a perceived human-mind-produced world.

The success in understanding of any scientific theory or publication is not only through the text, but additionally verbal expressions of the mathematical formulations and figures. Hence, the whole basic philosophy and working mechanism of any scientific work can be understood through the linguistic expressions, where there are not only crisp logic propositions, but most of the time vague, incomplete, uncertain statements that are more valuable in making further scientific developments. Such uncertain linguistic statements have fuzzy contents that can be assessed by fuzzy logic principles. Scientists treat figures as integral parts of their arguments, whose strength and soundness depend on visual representations as much as they do on linguistic representations. Arguments are expressed in terms of statements and this is one of the main reasons why the scientific philosophy has paid little attention to figures.

Fig. 5.1 Thinking gradients

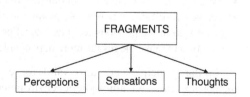

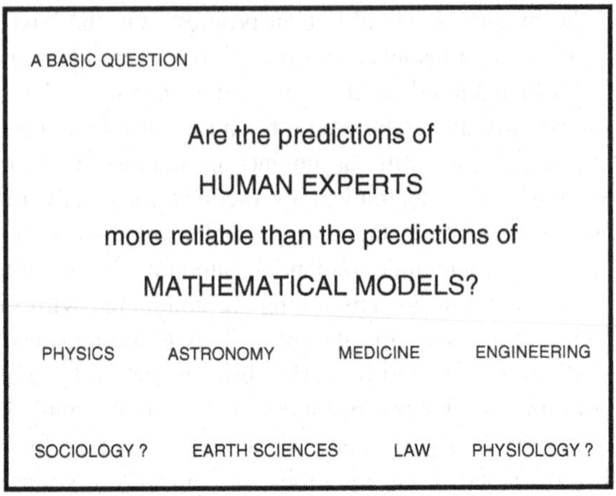

Fig. 5.2 Expert views versus
modeling

A BASIC QUESTION

Are the predictions of

HUMAN EXPERTS

more reliable than the predictions of

MATHEMATICAL MODELS?

PHYSICS   ASTRONOMY   MEDICINE   ENGINEERING

SOCIOLOGY ?   EARTH SCIENCES   LAW   PHYSIOLOGY ?

In everyday life human beings make many predictions especially on the basis of qualitative data and past experience. Additionally, expert opinions help to shape and refine such predictions besides the mutual discussion and confidence. In predictions there are similarities, which are the input information about the phenomenon concerned, output clues and the logical connectivity between these two sources of information. On the basis of certain clues, it is possible to make judgments about some target property, i.e. output information. The default of these judgments is the commonly available scientific thinking and its sublime version of logic (crisp or fuzzy) leading to rational results. This provides ability for any individual to develop actuarial models for various real-life prediction problems.

It is possible to make predictions either by crisp logic mathematical models or expert views. The major question that arises is presented in Fig. 5.2.

An important question is whether the predictions of human experts are more reliable than that of mathematical models? Experts make their predictions on the basis of the same evidence as for the mathematical foundations, but additionally they consider the usefulness of the linguistic data in the form of vague statements for the adjustment of the final model. Such vague information cannot be digested by crisp logic mathematical model, because any sort of uncertainty is defuzzified, i.e., rendered into crisp numerical forms. It is, therefore, expected that the fuzzy modeling by experts considering vague information is more successful than mathematical models, which are valid for ideal cases under the validity of a set of assumptions. Among the most important problems are natural phenomena predictions, because they have the following properties.

- Even the best mathematical models are not especially reliable.
- The best results seem reasonable predictions, but somewhat unsafe, and therefore as mentioned in the previous sections logical proportionality (directly or inversely) are imported to make the results more dependable.

Similar principles are also valid for engineering, geological, hydrological, meteorological, and atmospheric environmental phenomena. In order to move understanding towards a deeper and broader grasp of complexity, the emergent meanings need to be neither stable nor unstable, that is stable enough to rely upon them when generating hypotheses, concepts, and emotional attitudes, and unstable enough not to allow these concepts and attitudes to harden and become dogmas and addictions. In other words, after scientific thinking, meanings need to be fuzzy (flexible), ready to immediately respond to the changes continuously occurring in each of the countless dimensions of reality.

5.3 Science Philosophy and Engineering

Engineering services and structures are among the corner stones in any civilization establishment, growth and development for social harmony of the society. In the planning, design, construction, maintenance and operation of these structures, engineers had practical and creative artistic abilities in the past, but systematic education and training programs led them to get away from such abilities with more emphasis on the analytical and numerical ability improvements with deductive inferences leading to standard solutions. Although analytical intelligence and ability are indispensable gradients in mass production, their crisp and hard rule applications do not provide creative bases in engineering career. It is the main theme of this section to emphasize the significance of science philosophy entrance into the engineering education and training. Without such a basis engineers expect case study solutions and ready software or formulation matching in their problem solving stages, and hence, creative abilities are not cared for future improvements and advancements. Philosophy of science provides dynamism into the creative intelligence of engineers [9].

Engineers after their four year Bachelor of Science education, depending on their analytical thinking abilities, try to solve problems according to readily available formulations or software without creative thinking capability. This indicates that none of the formulations provide a unique solution of the problem at hand, but approximate results. This implies further that any formulation has improvement possibilities provided that the scientist or engineer wants to think analytically with the support of science philosophy. For an engineer philosophical thinking means to understand foundations of engineering problems not through the symbolic logic and symbols as in the formulations, but their logical rules [9].

In the past, master-apprentice training for an expert engineer has become more involved in the universities as if apprentice stage corresponds to Bachelor of Science, master stage to Master of Science, and finally, expert level can be viewed as the Philosophy of Doctorate. In these three stages of modern education system, "science" and "philosophy" should be emphasized even in engineering training. The graduates seem to have been empowered with analytical thinking capability, which helps engineers to memorize, transfer and ready use of knowledge according to past applications. It is emphasized in this chapter that science and its philosophical foundations can be given to engineers for better problem solving and even personal emotional and intuitive comfort, which help to improve practical and creative intellects. Engineers after their four year Bachelor of Science education, depending on their analytical thinking abilities, try to solve problems according to readily available formulations or software, without creative thinking capability.

5.4 Science, Scientisism and Engineering

Scientific researches bring forward frequently new theories or improvements of old ones. These theories enlighten the scientists on a scientific path after their scientific proof documentation. It is, therefore, necessary to filter each theory through the scientific tests in order to distinguish their features from non-scientific allegations such as scientism, conjurer, charlatan (quake), juggler, and illusionist. In the development of scientific ideas intuition, subjectivity and metaphysical thoughts can play triggering roles. This is due to the fact that a new, innovative and creative idea never lies in the medium part of a normal distribution (bell shaped curve) but they require deviations on either side so as to be extreme ideas that may rectify either the existing theories or completely replace them. The human thoughts can be quantified as a natural bell shaped (normal-Gaussian) distribution as in Fig. 5.3.

Such a distribution represents normal thoughts in the middle (as on the average), and the two tails include extreme thoughts, where the new scientific impacts lie in most frequently. As in this figure, most of the human thoughts and ideas are concentrated in the extensive middle range of the curve, where traditional, imitative, almost dogmatic, common sense information and knowledge exist including even uneducated people. The ideas in this range may help to get even academic promotions with not a scientific goal but rather title gaining task, which may fall in general into the domain of scientism coupled with dogmatic molds. This range may cover almost 95 % of the total area under the curve, which is equal to 100 %.

Fig. 5.3 Human thought distributions

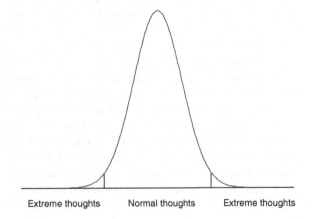

Extreme thoughts   Normal thoughts   Extreme thoughts

It is also possible to search for the scientific capability of a society in this area. The greater the middle area, the less scientific is the society, because there are classical information and knowledge common among the people without any interrogation on the scientific results and deductions. Such a society cheats itself as being scientific and enlightened social unit.

Evolution of scientific ideas, replacement with new developments and even those that have been exterminated continuously within the tails of the thought curve as extrasensory concepts, have revived leading to new scientific improvements. However, these are not metaphysical concepts only but also ways to the scientific generation channels with continuous discussions, debates and comments. These concepts are not necessarily objective and include partial subjectivity, because they are not yet ripe as scientific principles. In a way, science seeks a systematic information and knowledge network; it is not crisp as deterministic laws but includes vagueness, uncertainty, and incompleteness to a certain extent. Their origin lies within the uncertainty, metaphysical and extrasensory domains as sophisticated and complex roots but with time they give rise to better objectivity. Similar to a very rough sea with random waves, the scientific ideas and concepts are within such a domain, but science boat during its sail with rudder, engine, body and captain gets its firm navigation among the random ocean waves. Almost all of the scientific inferences come into existence down to the level of humankind's grasp from wild and complex media of thought through rational ripeness after the support of science philosophy and logical filtration. Scientific thought waves cannot be stopped completely but locally and partially they may be taken under control, which is the debatable end product of scientific theories.

In Fig. 5.3, the two-tail domains include extreme and even scientific anarchist thoughts without systematics, but rather in the form of linguistic (verbal) uncertain or fuzzy statements. These two tails are very useful for deduction of scientific inferences. The scientific developments move towards the middle range by time after understanding, and hence, dynamic scientific activities become as routine, classical and traditional. Those who share such scientific situations cannot be active scientist, but rather scientism starts to play role in academic promotions and perhaps classical scientific paper writings. This is what has been explained by Kuhn [5] as the normal (traditional) research path. Such paths may help to partial social enlightenment, but cannot provide its enlightenment for whole society. For instance, in the classical university education systems, if lecturers do not renew their knowledge and information then students take traditional education training. On the other hand, revolutionist scientific researches and activities appear in the two tails, which triggers (fires) new ideas with the falsifiability of the existing scientific knowledge. It is, therefore, possible to subdivide the curve in Fig. 5.3 into two mutually inclusive regions; those at the two extreme sides (tails) as revolutionist science and in the middle for normal science. Of course, the number of researchers in the tail regions is very small, whereas almost more than 98 % of the area is covered by the normal, classical, traditional and conventional scientific activities.

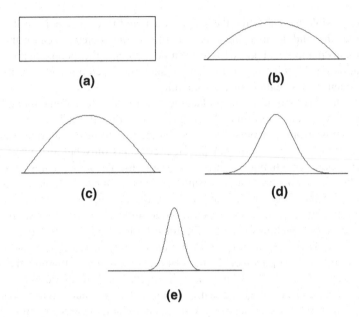

Fig. 5.4 Thought evolution

The evolution of scientific thoughts improves by time as in Fig. 5.4. Initially, as in Fig. 5.4a, almost all the spectrum of thoughts has the same level.

This is a theoretical case where each thought is given the same weight, which is naturally impossible. In such a society the ideas have not yet active position and dynamism; there is no systematic knowledge or information, where all known and unknown are mixed in a chaotic manner. Primitive societies have such a spectrum of scientific thoughts. With the initiation of scientific thinking some of the scientific knowledge appear as superior to others by triggering human mind, and therefore, similar information start to accumulate in a cluster, and hence, different information clusters appear each with different purpose and different levels of certainty. Such information can be reached after critical thinking and linguistic debates, which constitute the basis of science philosophy and then the ideas, are filtered through the logic. Such a mind activity gives rise to dominance of some regions in the spectrum domain, and consequently, Fig. 5.4a takes through scientific evolution the shape as in Fig. 5.4b. In this figure, middle range has dominance over the two tails, but the generative dynamo of scientific knowledge remains in the tails. This does not mean that the middle range cannot give rise to new ideas, of course, it does, but in a non-revolutionary manner as normal scientific developments. The number of researchers overwhelms in this range than the tails. Uncertainty remains in the two tails, but there is not a crisp boundary between the two regions (tail and middle range). One can imagine that theoretically the collection of scientific knowledge has the same value, whether in the metaphysical or positivistic domains, but there are conquests on the metaphysical domain so that the scientific knowledge and information increase with time. If one considers all the available knowledge (known or unknown so far) as 100 %, then as the scientific evolution increases the percentage of the scientific knowledge, the dark, i.e., unknown domain shrinks. Hence, as the knowledge becomes scientifically systematic, the influence range of the middle part also increases leading to a better enlightenment level in the society. The scientific alert of the society increases with the enlargement of the classical scientific knowledge dispersion within the society, and on the other hand, scientific dynamism also supports the same society for better science and technology levels by transferring knowledge from the tails towards the middle part. The bell-shaped curve in above discussions can be represented by two parameters similar to the Gaussian probability distribution function.

- As the scientific information increases the clusters during the scientific evolution also increases towards the middle range. This can be viewed as the average scientific level of a society similar to the arithmetic average parameter in the statistics terminology.
- Although scientific evolution causes concentration of knowledge in the middle range, but it never transfers unknown knowledge in the tails completely into the middle range. Hence, there is always a deviation from the average, which is similar to the standard deviation parameter in the statistics literature. There is a reverse relationship between the information content and the standard deviation, which is the concept used in many scientific activities concerning uncertain events [3].

Furthermore, Fig. 5.4c and d represent more evolutionary cases compared to Fig. 5.4a, in general, it is possible to consider Fig. 5.4e as a representative of today's information level in general. This is the shape that each society should strive to achieve for scientific and technological developments.

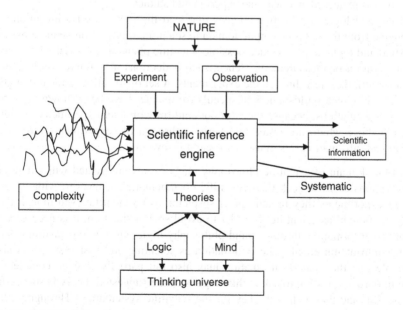

Fig. 5.5 Scientific inference engines

The scientific inference engine is given in Fig. 5.5, where the input thoughts are sophisticated, complex and very uncertain including metaphysics, imaginations or some of them.

For scientific inferences the general pattern shown in Fig. 5.5 is useful where the most significant component is the "scientific inference engine." The following points among the advices that may help for scientific information inference from the complexity world to systematic scientific information with the passages through the generalizations, predictions, assumptions, and debate stages.

- In cases of complex thoughts initially on the one hand observation, and on the other, mind and logic interpretation provide a preliminary grasp of the event under investigation leading to linguistic logical rules. At this stage, the scientist should visualize the events with his/her observations and thoughts by using mind, logic and preliminary information in the imaginary medium leading to scientific expectation forms based on logical verbal statements. If s/he can identify these verbal statements then they can be symbolized into mathematical expressions through proper notations. Consequences of such thoughts and scientific stages lead to scientific theorems about the phenomenon concerned. The validity of the theorems can be verified through different tests partially seeking for better representations. Even though the verification tests cannot be performed simultaneously, they can be postponed and delayed for convenient experimental setups. For instance, the deflection of light rays as foreseen by Einstein had to wait until Eddington had to inspect sun eclipse after many years, which led to observations and measurements for the verification and testing of the theory, and finally, it is decided that what Einstein had stated many years ago, was in accordance with the experimental studies.
- Mind and logic are the two couples that lead through the scientific inference engine from the unknown input domain to scientific information deductions.
- Mind and logic help to deduce possible scientific phenomenon even without any observation or measurement. Hence, such deductions may remain as imaginations until their test through the experiments. A set of so called rational thoughts from old Greek philosophers in speculative manner have been shown to be the only speculations, because they did not confirm with later experimental results. Only those which are tested and verified through the experiments remain as rational information sets in use with scientific evolution.

In the domain of scientific philosophy many events are treated with mind and observations so as to seek their scientific documentation. However, without testing or experiments they remain under suspicion as unscientific even though they may be thought as scientific. In order to avoid such a situation, it is necessary to treat the information through mind-philosophy-logic-observation quadruple and their confirmation enables the information as scientific and systematic. It is also possible that the treated information may also fail partially or even completely. The important question to ask at this stage is, if the suggested theory is successful in the test, and then is this the key for the scientific systematics? Throughout the historical background, there are different opinions on this matter. Shall we decide as scientific if any new theory passes the tests partially? Initial approaches based the scientific validity on the successful passage through the test and experiments for a given idea. Such an approach gives the impression that the scientific statements cannot be falsified, but confirmation through the tests is enough to verify its validity. This brings one to the conclusion that scientific information once verified, it cannot be falsifiable, which puts the science on a dogmatic statues. Popper [6] has suggested that only falsifiable theorems are scientific, because this principle leaves open door for future developments in the sense that one can bring better theory than the existing ones with even minor improvements. Falsifiability drives out all the dogmatic, traditional and classical static thinking procedures, but adds dynamism to the theorem, which has already been verified. This leads us to the path that all the theorems that have been verified up to now are falsifiable and one must not believe in scientific confirmations as a religion. This is a slight difference between a religion and science. This principle sheds light on the distinction between the religion and science. The falsifiability principle exposes all the scientific theories to criticism, which supports revolutionary scientific arena. The knowledge that is not falsifiable cannot be considered as scientific.

Starting from early times and today there are many theories that have entered into the scientific arsenal, but many of them cannot abide by the falsifiability principle. They have been classified as scientific under the light of verifiability. Especially, socio-economic and psychological events have been tried to enter the scientific arena according to verifiability principle. A simple example for this is the "individual phycology" theory by Adler [1], which is concerned with the spiritual insight of each individual. The principles of individual phycology have not been established according to scientific and experimental research and as he states thousands of individuals have been interrogated and the derived conclusions have been generalized exhaustively as scientific inferences. Unfortunately, many experts in the same discipline have relied on the verifiability principle of the scientific criterion and when they applied accepted general rules to different individuals they came across with invalid conclusions. They thought that according to the falsifiability principle of scientific measure, they can rectify the situation but they had always failures. An effective example for this situation is the case of a drowning man, who expects rescue operation. Here, the question is whether the man who runs for rescue behaves for the sake of saving the life or just for show off or to get the attraction of others as a brave man? On the other hand, another man who sees the drowning of someone and s/he does not try to rescue because s/he does not have self-confidence and reliance but has inferiority complex. It is possible to explain of both men according to scientific principles. Another example to this is the psycho-analysis by Freud [4]. In this case also without thinking the falsifiability principle, the conclusions are based on the verifiability approach, which fails in many cases. The reason of the verifiability principle arises from the empirical evidences, which is right as far as the scientific works are concerned, but the conclusions must not be subjective, they must be objective, which imply that even a single case of invalidity is sufficient for falsifiability in science. The objectivity in science can be obtained from the falsifiability principle only.

Falsifiability does not imply complete denial of the theory, because all the theories include some approximations or determinism on a set of assumptions. Especially, in the materialistic sciences such as physics, chemistry, mechanics, astronomy, etc. falsifiability does not imply that the theorem is invalid completely. Up to now, there are not scientific theories that have been falsified completely. This means that a falsifiable theory loses from its validity partially, so as to give ways to better theories or improvement of the same theory depending on the circumstances. Each falsifiable theory is succeeded by new or improved theory, which according to Kuhn [5] leads to revolutionist theories and scientific activities. So, falsifiability principle helps to improve the existing theories, whereas the verifiability approach renders each theory as if it is the final end without any further improvement. Falsifiable theory may need rectification by adopting many initial principles and the false sides can be amended. Search for completely new theories is a path towards the revolutionary knowledge and emergence of new ones. If one looks at the historical chronology, Ptolemy's earth centered universe mathematical formulations remain the same in the case of sun centered universe of Copernicus. This indicates that Ptolemy's theory is not wrong completely but under the light of new and recent information, observations, experiments and interpretations it is not sufficient for explanations. Hence, consideration of the false points in the Ptolemy's theory stimulated others and especially Copernicus through Nasr Al-Din Tusi [7] to arrive at a new theory with additional systematic considerations. Hence, invalid points in the Ptolemy's theory remained in the historical lines up to now.

In order to verify theories there are many different methods. Incomplete parts of the theory can be filled with valid and additional interpretations, which provide a patchy complement for the old theory, but with time new ideas and thoughts may integrate these patches into the whole system towards the perfection of the theory, but still each theory remains at the falsifiable state. For any theory to have evolution, it is necessary that there is a certain level of risk for its falsifiability. This implies that each theory has a certain level of risk; it must be tested with care and with every means its falsifiability must be controlled continuously. Hence, the falsifiability principle is like a fuel for dynamic and innovative new theories or suggestions of completely new research directions. If one sticks to verifiability principle, then the scientific theories enter into the belief world of a person with hindrance to further development and at the end such theories may take the form of religious beliefs. It is, therefore, necessary to keep in mind that every scientific statement, hypothesis or theory is falsifiable. In its historical path falsifiability has been accepted as suspicious cases.

5.5 Scientific Elements

Natural and social sciences are rather descriptive and there are always partially overlapping conclusions between different experts on the same phenomenon depending on their background and experience. Differences in opinion open ways of potential questions in training and research activities, which depend more on field or laboratory works where initially descriptive, rather vague, obscure and uncertain linguistic information emerge through observations, measurements and simple logical concepts. Vague and uncertain concepts are the basis for further discussions, because there is no systematic methodology for their assessment, control and acceptability by different parties. The best that can be done is the use of uncertainty techniques such as the probability, statistics and stochastic processes, but they require numerical data for the implementation.

Scientific consequences are dependent on premises that are logical proportions between cause and effect variables of the phenomena concerned. These proportions are verbal or linguistic statements, and therefore, at the initial philosophical thinking stage, they include vagueness and imprecision. As more scientific evidence becomes available rationally or empirically, either the validity degree of the statements increases or vagueness proportion decreases. Under the light of science philosophy, these statements are either assumed as absolutely correct or more frequently they have some uncertainty degrees. For uncertainty, objective probability attachment is a difficult task, and therefore, in practice, subjective (Bayesian) proportions can be attached to these statements. Apart from field or laboratory measurements that lead to numerical data, experience is the most precious and important piece of information, which are naturally linguistic in content. Natural and social sciences advance with accumulative and transitional experience to young generations. Experiences cannot be expressed by mathematical models, equations or algorithms, which are deterministic and depend on Aristotelian binary (crisp) logic. The crispness is rather unnatural for human reflections, and therefore, ignores the vague portion of information. Hence, the question is whether to render human thought into crisp logical principles or to change the logical propositions and inferences such that they account for vagueness. For instance, in engineering, classical physics and similar education systems, in order to get rid of the uncertainties, a set of assumptions are necessary prior to actual problem solving (see Fig. 5.6).

Idealizations in terms of assumptions including homogeneity, isotropy, uniformity, linearity, i.e., are among restrictions to natural events in their conceptual modeling. Without a set of assumptions, it is almost impossible to derive equations or mathematical models. Unfortunately, even today education institutions, especially in natural sciences, seek deterministic expressions with basic assumptions although the phenomena concerned evolve under uncertain environmental conditions. Many disciplines try to educate their members through classical and deterministic physical principles, mathematics, and statistical methodologies based on crisp logical principles (black and white). It is time to change this view with the treatment of uncertain data under the light of flexible logical principles. For this purpose, an innovative training should be founded on philosophical thinking and its systematic molding by linguistic logical (fuzzy) statements.

All our deeds in society, economy, administration, management, engineering, medicine and sciences take place in a complex world, where complexity arises

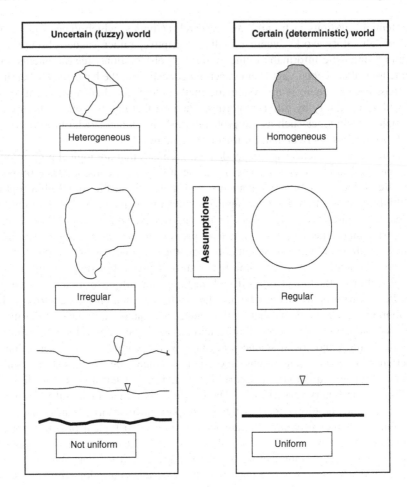

Fig. 5.6 Fuzziness versus determinism

generally from uncertainty in the forms of ambiguity. Humans address complex problems initially with ambiguity and ubiquitous features pervade most natural, social, technical, and economical problems. The only way for computers to deal with complex and ambiguous issues is through the fuzzy logic principles for systematic control, prediction and decision making.

The point of view taken in this book is to use innovative fuzzy logic principles in education, research and training systems of natural and social sciences and especially in engineering. This is the most needed tool in science, in order to relate linguistic (verbal) information and expert views first to logical statements and then to numerical data. In complex phenomena, fuzzy systems are very convenient for general and particular feature definitions. Fuzzy logic captures quantitative and empirical knowledge and provides simultaneous simulation of multiple processes and non-linear relations leading to soft computation techniques.

5.6 Modeling Principles and Philosophy in Engineering

Although independent from careers, there are thinking models that can be perceived easily after expert trainings. Constrains in such models vary according to dealing with natural or social events. Prior to the modeling of any complex event, a set of simplifying assumptions and idealizations help to imagine, visualize and construct preliminary model structure (Fig. 5.7). In any inductive or deductive modeling, input (cause) or output (result) variables are considered for their mutual associations, relations or dependence. Consideration of input and output variables with their mutual relationships provides transformation system of inputs to outputs simply in the form of "black-box" modeling, which implies that the physical features of the transformation system are not well-known. Such models are rather restrictive and not valid universally for temporal and/or spatial variation description of the phenomenon concerned. In any simple model, engineer wants to identify the transformation mechanism so as to estimate or predict the output variable from a given set of input data (Fig. 5.7).

In such a modeling structure there are three different components that can be used in practical applications.

- If apart from the input other two components are known then it is "filtration" (smoothening) process of the input data.
- In case that model input and output variables are known and the modeling problem is to know the transformation mechanism then it is "identification" type of modeling process, which is most frequently used in practical applications. This is the most difficult modeling type and many formulations, equations and algorithms fall into this category.
- If the input and the transformation mechanism of a model are known then the type of modeling is "prediction" or "estimation" process. It is the most frequently used model by engineers and many others.

It is not possible to model any phenomenon without making a set of assumptions, which are not concerned only with the structure of the model but also with the input and output variable properties. For rational assumption allocations one should consider not only mathematical complications but their geometrical imagination, visualization, and finally, a convenient design (Chap. 4).

Similar to evolution of human and social sciences, engineering thoughts and methodologies have their evolution patterns leading to continuous, accumulative and sustainable development. In some societies engineering is thought as a career

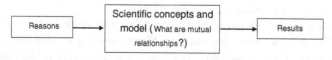

Fig. 5.7 Black-box modeling

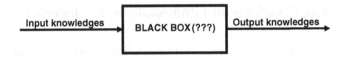

Fig. 5.8 Black-box models

that does not have any connection with philosophy. Such a trend led to engineering understanding as intake of even basic information and knowledge blindly through dogmatic memorizations without questioning and critical review. This is referred to as "black-box" model, descriptively given in Fig. 5.8. These models are philosophical externally but internally there are no philosophic aspects.

These models do not care for the internal physical relationships between input and output variables, but try to fit the most suitable mathematical, probabilistic, statistical, stochastic or any other model including artificial neural networks for transformation mechanism. They do not care for physical relationships and the sole purpose is to match output variable pattern to input patterns only without any philosophical thoughts or principles. They are for saving of current situation in the simplest and cheapest manner and they do not represent the real situation about the phenomenon concerned. Even though black-box models are used frequently in engineering, they are without internal scientific documentations, and hence, they are abstract in a way and provide approximations only [8]. Nonexistence of philosophical principles in the black-box models does not mean that they do not have logical bases. Valid logical rules are only for relationships between the input and output variables without touching to the internal structure (generation, transformation) mechanism.

A significant example for such a simple modeling is the relationship between the stress and deformation, which is referred to as the Hooke's law in engineering. Although it is referred to as law, it provides the simplest model that combines deformation, ε, input to stress, σ, output. The basic assumption in this model is that the relationship between the input and output variables has proportional linearity, which implies that as the deformation increases, stress increases. This law can be written symbolically in a mathematical form as,

$$\sigma = E\varepsilon \tag{5.1}$$

where E is the proportionality coefficient, which is referred to as the elasticity modulus in engineering. Any classically educated engineer will say that E is the elasticity modulus without any further explanation. This indicates that classical engineering education without philosophy of engineering and logical rule inferences leads to ready information storage in the memory and their mechanical use at times of request. However, even an engineer with some philosophical thinking background can make a series of interpretations (Chap. 5).

Even non-specialists in engineering accept that without deformation (stress), it is not possible to have stress (deformation). Hence, s/he implies that there is a relationship between the two variables. The next question is what is the form of the relationship between the two? Rational thinking leads to easily that it is a directly

Fig. 5.9 Stress-deformation
relationships

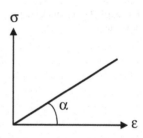

proportional relationship. Still there is another question, what is the type of relationship, is it linear or non-linear? Perhaps this may not be answered quite easily, but as a
first approximation one is able to make the simplest assumption of linearity.

In the Hooke's expression given in Eq. (5.1), anybody with free thinking may suggest the following points as his/her interpretation according to personal capability.

- From the mathematical point of view s/he may name E as the proportionality
  constant. Additionally, consideration of very small variation intervals (dσ and
  dε) it is possible to notice that E = dσ/dε, which implies that the constant can
  be defined as the ratio of stress to deformation or it is the derivative of the stress
  with respect to deformation. Furthermore, the same constant is equal to the
  slope of the linear relationship between stress and deformation.
- From the engineering point of view, simply E = dσ/dε implies that the same
  amount of stress variation corresponding to per unit variation in the deformation,
- A closer look at the relationship indicates that E is the measure of correlation
  between σ and ε, as E = tanα, where α is the slope of stress-deformation graph
  straight-line (see Fig. 5.9).

After all one can feel that the naming of E as the elasticity modulus is an engineering convenience, because it is related to the property of the material. However,
if an engineer knows that E is the elasticity modulus, without the aforementioned
points, then s/he cannot question capability and philosophy of engineering principles.

Under the light of what have been explained above, it is possible to reach simple models and relationships between two variables through a constant on the
assumption that the relationship is linear. Such a thought leads to simple models and the literature is full of such relationships in the forms of laws as Newton
second law in physics, Ohm law in electrical engineering, Darcy law in groundwater researches, etc. Another point is that in all the linear models the simplest
geometric shape, straight-line (linearity) is considered. A rational interpretation of
this point is that in such simple and regular shapes one implies that the material is
homogeneous and isotropic. Of course, such assumptions will be approximately
valid in any practical application.

After the explanation of the elasticity modulus, again through critical thoughts
and questions one can think that if there are different materials such as plumb,
iron and steel their comparisons give way to another debate as to which one has
the biggest (smallest) elasticity modulus? In such a comparison, numerical values

Fig. 5.10 Different elasticity
modulus

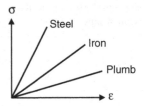

are not sought but nominal (verbal) quantities play role. Since elasticity modulus expresses stress variation per unit deformation, one can reason as to which one of these three materials will be subjected to the biggest stress under the same deformation amounts. It is possible to question, are smooth or hard materials more deformable under the same stress values, or vice versa? The rational answer is that even without engineering specialization, hard materials require more stress for the same deformation. This leads to the logical conclusion that hard materials resist to stress more, and hence, their elasticity modulus will be comparatively bigger than smooth ones, and consequently, they are comparatively more durable. After all these facts, one can imagine, visualize and immediately draw the following graph in his/her mind (Fig. 5.10).

Hence, without any number and even formulation rationally thinking mind can draw all possible rational information and knowledge as explained above. Of course, s/he uses philosophy of engineering and logical information as one senses, but in the case of philosophy of engineering fundamentals and logical principles such minds enrich the discussion and an objective road is paved towards more quantitative, symbolic, mathematical and numerical solutions. All the conclusions so far did not require any numerical work, but in any engineering application the numerical quantities are required so as to make numerical calculations for the solution of the problem at hand.

Neither mind, nor philosophy of engineering nor logical principles says something about the numerical quantities. The only way open towards this direction is the experimentation and measurements either in the laboratory or in the field. As for the above example, rational thinking cannot identify the numerical values of elasticity modulus for plumb, iron or steel. Hence, it is very important that prior to any experimentation or numerical calculation, it is necessary to identify the representative model with linguistic (verbal) information derivation through plain rational thinking, philosophy of engineering and logical principles. Subsequently, after making the linguistic background available then one can perform experimentation. Especially in engineering, these stages are important, because an engineer should design on the paper as a model what s/he imagines in the mind. For instance, an architect or engineer should draw on a piece of paper the geometry of what s/he thinks about the problem and then by critical discussion s/he should criticize his/her suggestions linguistically so as to reach a final shape (design) among several alternatives. All what have been explained so far are without entrance into the black-box model.

After all the aforementioned discussions, in order to identify inside functions of the black-box, it is necessary to explore the internal generation mechanism through a series of relevant questions. How it works? Why it functions in this manner? (see Fig. 5.7). In this way, internal philosophical principles start to take over for detailed rational, philosophical and logical interpretations and conclusions. Answers are sought for each one of these questions, so that the black box starts to become grey, but it never reaches to completely white-box (every single detail is known) case.

What are the variations in the material body in transforming stress to deformation? How materials behave? For rational and logical answers, internal relationships become important in engineering discussions. Answer to such questions is concern of scientists rather than engineers. For the answers, it is necessary to know or imagine the internal structure of the material as minerals' positions, molecular structures, and their reaction behaviors against the stress, which are of scientists' concern and they should make a set of assumptions and simplifications to arrive at some practically usable conclusions. Of course, they will need to get support from the philosophy of science, which should have some ingredients in the engineering domain also. For instance, is it possible that subject to force, dimensions of the material, and hence, the minerals shrink, but do they retreat to their original form after the force release? If such a situation is valid, then the researchers may conclude that the material can change its shape elastically. Hence, such a material can be used in any engineering construction; otherwise plastic material usage is dangerous right from the beginning.

In engineering most often black-box models are widely used and they provide practical developments rather than scientific. In grey box models, scientific and engineering modes take place simultaneously, and hence, engineers benefit from scientific outputs and scientists may also benefit from engineering design and practical productions for sensible instruments, which help to support scientific investigations and developments. In any grey-box model, there is a common area of interest from the scientific and engineering points of view as in Fig. 5.11.

Philosophy of engineering empowers engineer with scientific views in problem solutions through questioning not only the designs but also formulations, equations, software and algorithms as to what are their internal functions?, and how do they work? If an engineer is capable to deal with such questions, then s/he is also able not to depend blindly on the ready formulations, but to modify them partly or to suggest a completely new formulation or methodology. Although scientific solutions are universal, engineering solutions benefiting from scientific results may require local modifications. Such modifications cannot be succeeded by an

Fig. 5.11 Science-engineering relationships

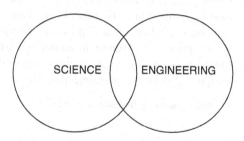

engineer who sticks to traditional and classical education information and knowledge. S/he should question any applicable plan, project or implementation according to temporal, spatial and environmental conditions. Such adaptations cannot be achieved if engineer does not care for philosophy of engineering together with logical principles linguistically prior to any type of numerical quantification.

An engineer empowered with philosophy of engineering cannot accept the mechanical outputs from available formulation, equation, and algorithm, and especially, in our days, ready software is the main nuisances in engineering applications, if the internal philosophical and especially logical structures are not well appreciated. Since science, and especially, engineering solutions are approximations, engineer should keep in mind that for the same problem, there may be alternative solutions, and therefore, s/he should try and reach to a set or few solutions by keeping in mind the philosophy of engineering principles and then again in the same manner the best and convenient optimum solution can be identified after an effective decision making. In order to find the most suitable solution for the purpose, the basic theoretical foundations must be appreciated to a certain extent by engineers.

Recently, technologic innovations reached to the extent that almost every day there is an innovation and people prefer to buy the latest technological improvements. In these innovations, the role of engineers cannot be denied. In any product ethical, aesthetical, artistic and cultural values are also important. Engineers, instead of being individuals, who apply scientific conclusions, should also have knowledge in social, ethical, cultural, economic, etc. aspects also. Combination of these interdisciplinary topics is possible through philosophical and logical principles, and hence, philosophy of engineering provides an additional wing to engineers to deal not only with technological developments, but also share in any basic scientific research for further innovative developments. Philosophy of engineering bridges between the engineering application approaches by giving them a scientific content, which aids engineers to criticize, discuss and try to reach the best solution under the light of currently available knowledge. Interaction between engineering and philosophy is not for practical applications through internal ambitions but more through their understanding, explanation, solution and production.

5.7 Model Criticisms in Engineering

Unfortunately, in many cases after the completion of any engineering task through modeling, there are not critical criticisms about the methodology, formulation, algorithm, mathematical expressions, formulations, principles and assumptions neither prior nor posterior to modeling. Although there are many examples, the following points are some citations, where the expressions within the parenthesis are ignored in everyday engineering life.

- Safety factor (ignorance coefficient);

- Other coefficients (elasticity modulus, runoff coefficient, porosity, permeability, etc. are not real, but mostly virtual or fictive!);
- Software (philosophic and logic flow diagram!);
- Assumptions are not valid at all times and places (validity!);
- Science is verifiable (Falsifiable!);
- Infinity (possible!);
- Point has zero dimensions (How? One can still see it!);
- Force is equal to multiplication of mass by acceleration (conditions!);
- Hooke, Newton, Ohm, etc. laws are different from each other (Different symbols, but the same proposition for two variables!).

One can make plans, calculations and projects without questioning these points by accepting them as valid under any circumstance. Engineers, academicians, administrations, old graduates of engineering may also stick to the same assumptions. They may think that without critical criticism, comment or discussion and without any doubt their ship sails in the ocean with no problem and they also transfer such static knowledge to other specialists. Conventional and classical engineers may gain even respect for doing so. If engineering is to apply whatever s/he may be thought during education through the memorization and classical applications, then the static knowledge are applied rather blindly without any discomfort. Is it not better to question them with doubt (skepticism) and rational thinking? Is there any harm in doing so? However, those who have been robotized with the knowledge in their memory might not feel comfortable. Any society empowered with critical thinking and philosophy will advance towards better and more illuminating future. The ratio of educated people may not be very significant from knowledge production point of view, but if education empowers these individuals then the knowledge storage and dynamic generation process will keep the society alive, generative and active in many deeds. Through the critical comments, discussions and thinking any society will possess engineers, doctors, managers, etc. who can give rise to additional and new information according to their ability in problem solving. Even though there may be many engineers who would like to stick to static information and knowledge but others will be ambiguous to activate the thought system for further productive and new ideas. Traditional engineers are in majority in any country, but they can be stimulated through the critical discussions once they are aware of the philosophy of engineering principles.

Now, let us start to criticize all the points that are given above list. Safety factor is a kind of mistrust in engineering calculations and even in scientific formulations, because after the application of these findings and calculations prior to design and structural dimensioning, engineers are advised to augment final numerical result by multiplying it by a number greater than 1 and mostly by 1.5. This is tantamount to saying that engineer does not trust the scientific findings and to be on the safe side s/he increases the result by about 50 %, which brings extra costs and work. Such a safety factor is due to the suspicion of engineer concerning the verification and correctness of scientific formulations. Such a suspicion is useful, but it does not help to augment engineering knowledge. Safety factor serves for a safer structural design, but engineering thinking becomes more static, because s/he does not question the methodology or formulation so as to improve its better validity. It would be better to improve the methodology by critical questioning rather than using a global factor such as the safety value. Otherwise, classical engineers accept that safety factor cannot be criticized and it must remain as is. By assuming that the safety factor cannot be criticized, one closes the door for the improvement of productive engineering thoughts. This also cuts the road for further idea generations rationally and experimentally, thus leaves engineers at the door of ignorance. Criticism of this factor from scientific knowledge point of view opens the door of thought experiments leading to better understanding of the phenomenon under study. Engineer then realizes that the construction material cannot be homogeneous, isotropic, linear, etc. and each one of these points leaves an unknown effect partially in the overall phenomenon; hence, s/he thinks that the safety factor covers all these uncertainties in a safe manner. Recently probability, statistics, stochastic, and similar uncertainty methodologies solve the same problem on the basis of risk (reliability) concept (Benjamin and Cornell 1975).

On the second line in the list above, are some other parameters that should be criticized rationally and logically by considering philosophy principles. One can realize that they are not really constant, but in practical studies after much experimental work they are adapted as the arithmetic average values representing the phenomenon under investigation. Furthermore, any equation and formulation used in engineering are in the forms of arithmetical averages, and this is the reason why each time one cannot have the same response in reality, but engineers and scientists alike after a set of assumptions rendered the uncertain forms into certainty domain through averages. In engineering not only arithmetic averages, but depending on the situation, geometric averages should be used. Averages imply that in any engineering calculation there are errors always. Rationality, logic and philosophy in engineering empower engineers towards the reduction of such errors as much as possible. Engineer must keep in mind that there are possibilities of refinement and such a search is possible if the engineer has philosophical thinking ability.

On the other hand, today much ready software is used in engineering works, unfortunately, again blindly as a black-box model, where the internal philosophical, logical, rational and physical functions are not known sufficiently or not cared for. Software may have safety factor or similar assumptions in its internal structure. Any software has philosophical and especially logical propositions and without being aware of these features the programmer cannot write any computer program. Additionally, even complex software has step by step logical statements and formulations in its body. Engineer must not use software about any topic without knowing at least the internal philosophy in terms of flow diagrams and logical statements. Otherwise, engineer cannot appreciate the outputs from software, and consequently, their interpretations, critical criticisms and comments are not possible. Such basic missing does not give opportunity to engineer to write reports. Without philosophy of engineering and logical rules, software is like a toy of a child, who does not care about its production, but uses it blindly.

5.8 Rationality Matrix

Concepts, terms, definitions, propositions and inferences are expressed linguistically. They can also be symbolized through the symbolic logic principles. Herein, a rational reasoning matrix will be explained for modeling on the bases of mind experiments and logical principles with philosophical ingredients. For this purpose, all the input and output variables in any modeling will be shown by small letters. Variables are indicated, in general by x, y, z, u, v, s, t, etc. The first question for the investigation of the event is whether there are relevant rational relationships among input and output variable couples? If there is, then the next question is what are the symbolic relationships? How do they vary versus each other functionally? It is possible to identify the relationship possibilities rationally. To do this, one can prepare a table including input plus output variables number, n, in the form of n × n dimensional matrix. Systematically, it is in the form of n × n square matrix. For instance, in case of four variables (three inputs, v, x, y and a single output, z), then the matrix takes its shape as in Fig. 5.12, where the first row and column have the same symbolic variables. In other rows and columns these variable pares have special pigeon holes.

One can write down the properties of such a rationality matrix by considering possible rational relationships.

(1) The total number of input and output variables defines the dimension of the rational matrix. If variable number is n then the matrix dimension is n × n.
(2) Apart from the first row and the column in any row-column intersection, two variables have a common rational property.
(3) Since along the main diagonal each variable is paired with itself, there is a complete relationship, hence each pigeon hole is attributed by letter C.
(4) The rational matrix is symmetrical with respect to the main diagonal, because the relationship between x and y is the same for y and x. It is, therefore, enough to consider the upper triangular section only.

Fig. 5.12 Rational matrix

(5) Each pigeon hole should have either directly proportional relationship indicator as ↗; inversely proportional as ↘; complete relationship as C; or rationally inde-finable relationship, ?. In the last case, rational thinking cannot specify any rela-tionship between the two variables, and hence, it is necessary to try and find the relationship through experiments or available numerical data.

By considering these points, all the pigeon holes are filled with convenient symbols and its completion guides to possible global relationship between the input and output variables. In the interpretation of such a matrix, the following points must be taken into consideration.

(1) In the rationality matrix, mind is not capable to provide type of relationship at pigeon holes with ? sign. It is necessary either to have observations and/or measurements for final (directly or inversely proportional relationship) deci-sion at these locations. For instance, in Fig. 5.12, it is not possible to decide rationally about the relationship between x and y. It is necessary to have data on these two variables in order to be able to identify the relationship between the two variables through the scatter diagrams. According to the visual rela-tionship between the two variables the scatter diagram will indicate either ↗ or ↘ sign. Hence, the rational matrix locations will be filled in in a complete manner as in Fig. 5.13.

(2) It must not be forgotten that in general directly or inversely proportional rela-tionships are considered linear frequently. The mind cannot decide easily about the non-linearity of the relationships.

(3) After the completion of the rational matrix, one can make different interpreta-tions. For instance, if the question is about the type of relationship as subject v variable in terms of others, then this variable, as can be seen from Fig. 5.13, is dependent on the others. As x increases v decreases; as y increases v increases; and finally, as z increases v also increases. If any one of these vari-ables is thought as subject then similar interpretations can be made.

(4) It is also possible from the rationality matrix to deduce the type of functional relationship. For instance, in mathematics the relationship between v, x, y and z variables can be written implicitly as $v = f(x, y, z)$. In engineering, such

Fig. 5.13 Complete rational matrixes

	v	x	y	z
v	T	↘	↗	↗
x	↘	T	↘	↘
y	↗	↘	T	↗
z	↗	↘	↗	T

implicit expressions do not have any practical meaning. It is, therefore, necessary to try and find explicit relationship between the variables. By consideration of all what have been said about the relationship from the rationality matrix in Fig. 5.13, one can write down the following alternatives,

$$v = a\,(yz)\,/x$$
$$v = a\,(y+z)\,/x$$

etc. Herein, a is a constant parameter that can be obtained only from experimental studies through the measurements, i.e. data.

In order to be able to decide which one of these alternatives is valid, it is necessary to benefit from basic and simple rules. The first information that one can use is to check the units of both sides. In any rational deduction the units must be the same. For instance, if the unit of the left hand side is in m, kg or Watt then the combined unit on the right hand side must be the same, m, kg or Watt, respectively. On the other hand, prior to the unit homogeneity, the initial and boundary conditions of the physical event may help to identify the valid equation. In order to shed light on this point let us consider the power of a pump as follows.

Let us try to obtain a formulation for the power of a pump that will haul water from a lower to a higher level. Let the pump power, P and the elevation height, h. Logically, the power will be dependent on several variables, but the most important ones are the amount of water (discharge), Q, and the density of water, γ. There are four variables (P, h, Q, γ) and the dimension of the rational matrix will be 4 × 4 as in Fig. 5.14.

Rational thinking will indicate that there are directly proportional relationships between power and other variables. Hence, P as dependent (subject) and others independent variables, the implicit formulation can be shown as P = f (h, Q, γ). One can write exhaustively explicit form alternatives as follows.

$$P = h + Q + \gamma$$
$$P = hQ + \gamma$$
$$P = h\gamma + Q$$
$$P = h + Q\gamma$$

or

$$P = \gamma hQ$$

Fig. 5.14 Pump power rational matrixes

	P	h	Q	γ
P	C			
h		C		
Q			C	
γ	↗	↗	↗	C

Each one of these expressions satisfies directly proportionality principles as in Fig. 5.14. However, there is only one among them that is valid from all logical, rational and unit homogeneity points of view. How can one identify this expression? In order to answer, one can look at the physical conditions of water pumping as well as unit homogeneity of both sides. Consideration of the physical initial and boundary conditions of the physical phenomenon indicates that the last expression is the sought alternative, which satisfies all the requirements in the rationality matrix (see Fig. 5.14). The first expressions can be tested as, say, if h = 0 then P ≠ 0, and hence, there is physical implausibility. How is it possible to have pump power, when there is no elevation difference as height? Similar reasoning eliminates other alternatives except the last one. Furthermore, the other alternatives provide the result of adding apples with pears, which is not possible. Detailed explanation of the rationality matrix is given by Şen [8].

After what has been explained in this section, one can say that engineer should have similar logical and rational reasoning so as to grasp the background of the equations that s/he uses. If this is achieved, then the engineer is ready to take any action in front of any problem and such logical and rational reasoning also provide the principles of software writings.

5.9 Mathematical Functions and Their Translations to English Language

During the secondary school and university training, the students have learned names of various mathematical functions and they also executed rather mechanistically different mathematical procedures such as integration, differentiations, and alike on these functions. Unfortunately, during the mathematical operations rationality, criticism, doubt and the logical principles as the bases of such operations have not been toughed but rather memorized knowledge have been put into functional procedural forms without understanding physically or practically what is made. This has led the engineers to rather dogmatic knowledge sources with transference knowledge without rational grasps. The author realizes that for scientific research and development any knowledge must be taken into consideration even with approximate reasoning. Many engineers know by heart that memorization cannot be helpful for generation of new knowledge, information or know-how. Generally, what we encounter in practical life as functional features are summarized in Fig. 5.15. This figure includes only very frequently used functions in engineering activities; and it is not a complete list of functions.

Linear equation. It implies that the logical relationship such that at least two variables (x, y) has directly proportional and linear properties, which can be expressed mathematically as,

$$y = i + sx \tag{5.2}$$

Fig. 5.15 Mathematical function forms

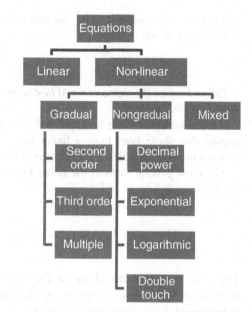

Fig. 5.16 Linear lines (geometry, function)

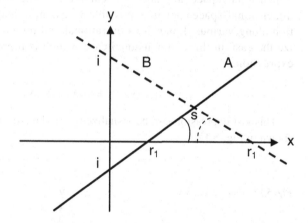

Its geometric form (function, straight line) is given in Fig. 5.16. In this expression x (y) are referred to as the input (output) variable in the modeling methodology; causal (resultant) variable in philosophy; independent (dependent) variable in mathematics; and predictand (prediction) variables in engineering.

There are two constants in the linear equations, which are intercept, i, and the slope, s. Intercept corresponds to a constant value on the y axis for x = 0. The slope can be regarded as the increase in the y value for unit increment in the x value, which is tantamount to saying that s = dy/dx, i.e. derivative of y with respect to x. Moreover, Eq. (5.2) has a single root, r_1, which corresponds to a constant value this time on the horizontal axis for y = 0. This last condition yields $r_1 = -i/s$. Alternatively, one can rewrite Eq. (5.2) by taking into consideration the root as,

$$y = i(x - r_1) \qquad (5.3)$$

Depending on the value of $s > 0$ ($s < 0$) the geometric shape (function) appears as A (B) as in Fig. 5.16.

All scientific laws can be expressed with Eq. (5.2) if $i = 0$ and $s > 0$, which implies that these laws have a linear line shape implying that the relationship between the two variables is in the form of directly proportionality.

The most general form of the linear line can be written with n different inputs $(x_1, x_2, x_3, \ldots, x_n)$ as,

$$y = i + s_{x1}x_1 + s_{x2}x_2 + s_{x3}x_3 + \cdots\cdots + s_{xn}x_n = k + \sum_{i=1}^{n} s_{xi}x_i \qquad (5.4)$$

where s_{xi} indicates the share of the i-th variable among the sloppy linear surface in n dimensional space. If this expression is considered with two independent (causative, input) variables (x_1 and x_2) then it shows in three dimensional space a plane. We cannot imagine linear hyper surface in n dimensional space, but it is possible to examine and deduce many information along various dimensions by rational and logical thinking. Let us visualize the case in three dimensional case, which is represented by the following expression.

$$y = i + s_{x1}x_1 + s_{x2}x_2 \qquad (5.5)$$

This last expression can be visualized as a plane in three dimensional space as shown in Fig. 5.17.

Fig. 5.17 Plane geometry

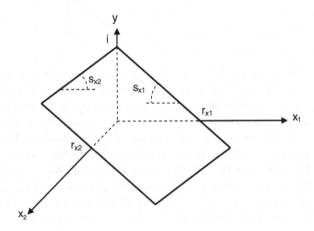

Additionally, Eq. (5.5) has two roots, one on the x_1 and the other x_2 axes. It is also possible to rewrite this equation in terms of the two roots, $r_{x1} = -i/s_{x1}$ and $r_{x2} = -i/s_{x2}$ (see Fig. 5.17) as,

$$y = i\,[(x_1 - r_1) + (x_2 - r_2)] \tag{5.6}$$

Second degree curve (Parabola). Addition of a second order term to the linear line in Eq. (5.2) leads to second degree function, which is called as parabola (Fig. 5.18).

$$y = i + s_1 x + s_2 x^2 \tag{5.7}$$

This equation has two roots, r_1 and r_2, on the horizontal axis. In terms of the roots one can write the equation as follows.

$$y = a\,(x - r_1)\,(x - r_2) \tag{5.8}$$

Herein, the constant a linguistically implies peak, P, (maximum), or valley, V, (minimum) points along the curve. In Fig. 5.18 A (B) curves are represented by $a > 0$ ($a < 0$). In each one of these curves, there is a single peak or valley. At any one of these two points there is a horizontal tangent. As a basic definition derivative means slope at these points $dy/dx = 0$. Simply derivation of Eq. (5.7) yields,

$$\frac{dy}{dx} = s_1 + 2s_2 x \tag{5.9}$$

By equating this expression to zero, it is possible to find the abscissa of the peak (valley) point projection value on the horizontal axis. This abscissa is equal to $x_1 = -s_1/2s_2$ as in Fig. 5.18). The positive or negative value of x_1 gives information about the peak and valley situations. It is simply possible to transform this last

Fig. 5.18 Second degree curves

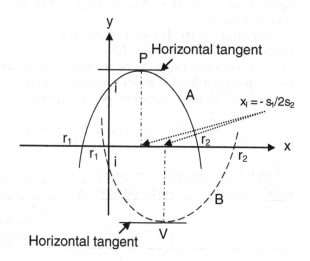

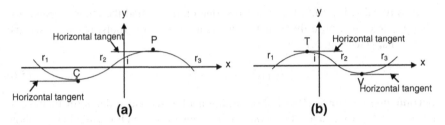

Fig. 5.19   Third degree curves

expression in Eq. (5.9) into a linear form as in Eq. (5.2). For this purpose let us denote the derivative by $t = dy/dx$, and then,

$$t = r_1 + 2r_2x \qquad (5.10)$$

Comparison of this expression with Eq. (5.2) provides all the linguistically interpretations for s_1 and s_2.

Third degree curve (Cubic). Addition of another term with the next integer power to Eq. (5.7) appears mathematically as,

$$y = i + s_1x + s_2x^2 + s_3x^3 \qquad (5.11)$$

Figure 5.19 exposes all the notations in this expression and its geometry (functional form). In general, there are three roots and accordingly the root inclusive form of the same expression can be written as follows.

$$y = a(x - r_1)(x - r_2)(x - r_3) \qquad (5.12)$$

Herein, a is the shape factor and depending on its positive or negative value the geometrical form of the expression is given in Fig. 5.19a and b. According to this figure, there are two turning points as peak (valley) and valley (peak) along the trace of the curve.

Comparison of this last figure with the previous two cases (linear and parabola), one can deduce the rule in Table 5.1.

Additional information can be obtained from the turning point concept. Since turning point is defined as points with derivative equal to zero, accordingly derivative of Eq. (5.11) leads to a second degree equation,

$$t = s_1 + 2s_2x + 3s_3x^2 \qquad (5.13)$$

Table 5.1   Degree-turning point relationship

Degree of equation	Number of turning point	Linearization derivation degree
1	0	0
2	1	First derivative
3	2	Second derivative

It is possible to make similar interpretation by comparison with the second degree curve as explained above. Accordingly, one may give meanings for each slope, s_1, s_2 and s_3, which is left to the reader. Another point for the readers is the second order derivative expression of Eq. (5.13), which is,

$$u = 2s_2 + 6s_3x \tag{5.14}$$

Herein, u is the derivative of the derivative d(dy/dx)/dx. It is now possible to compare this expression with the linear case (see Eq. 5.2), and consequently, relevant meanings can be given for s_2 and s_3 in the u-space. Another rule that can be deduced from these discussions is that in mathematics provided that the powers are integers, one can reduce the basic equation into a linear form by taking successive derivatives.

Multiple degree curve (Polynomial). The most general form of all the previous expressions can be represented up to n-th degree integer valued power equation as,

$$y = i + s_{x1}x + s_{x2}x^2 + s_{x3}x^3 + \cdots + s_{xn}x^n \tag{5.15}$$

By consideration of the Table 5.1 one can deduce that the number of turning points is equal to the degree of the multinomial minus one. This implies that in a multiple degree integer power, n, curve geometry (function), the number of turning points is equal to $n - 1$. Let us consider 5-th degree multiple degree equation as,

$$y = i + s_{x1}x + s_{x2}x^2 + s_{x3}x^3 + s_{x4}x^4 + s_{x5}x^5 \tag{5.16}$$

One can rewrite equivalently in terms of roots as,

$$y = a(x - r_1)(x - r_2)(x - r_3)(x - r_4)(x - r_5) \tag{5.17}$$

The reader can think rendering this expression into a linear form after successive derivative operations. For a reader who does such a task, s/he will be able to interpret such expressions on philosophical bases linguistically and transformation of symbolic logical mathematical expressions to verbal expressions and vice versa will not be difficult for him/her. Such a task melts almost all the difficulties through philosophical, epistemological and linguistic wordings and s/he can write down these wordings and logical sentences (propositions) into mathematical symbols.

Fractional power curve (Power function). In practical engineering studies, most often equations similar to the parabola, but with decimal power are encountered and such expressions are widely used in various topics. In the following the simplest form of the power equation is written.

$$y = ax^n \tag{5.18}$$

A set of curves (n > 2 and n < 2) derived from this expression is shown in Fig. 5.20 for n = 2. Compared to the case of n = 2 for n > 2 (n < 2) weaker (stronger) slopy curves emerge in the form of a family. The coefficient a is referred to as the scale parameter, whereas n can be referred to as the shape parameter.

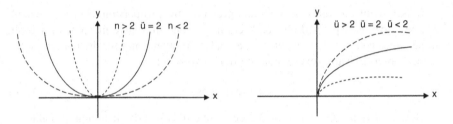

Fig. 5.20 Power functions

The linearization of this expression is not possible by successive derivatives, because the power is not an integer. For linearization of Eq. (5.18) the simplest way is to define a new variable as $z = x^n$ and hence $y = au$ takes a linear form. However, such a transformation does not provide any information about the meaning of n.

On the other hand, taking the logarithm of both sides in Eq. (5.18) provides another engineering clue of linearization as follows.

$$\log y = \log a + n \log x \qquad (5.19)$$

One can now define the logarithm of y on the left hand side as a new variable $u = \log y$; the scale logarithm as $S = \log a$; and finally the independent variable logarithm as $v = \log x$ then the linearized form appears as follows.

$$u = S + nv \qquad (5.20)$$

Comparison of this last expression with Eq. (5.2) indicates completion of the linearization process, where with new variables for $u = 0$, Eq. (5.20) yields $n = -S/v$ or in terms of logarithm $n = -\log a / \log x$ or $S = -n \log x$.

If one wants to develop a simple practical rule, it is useful to reconsider Eq. (5.18). Instead of (x, y) domain, if (logx, logy) domain is taken into consideration then one can obtain a linear line with intercept on the horizontal axis, which is equal to $n = \log a / \log x$. This indicates that taking the logarithmic axes instead of natural ones, opens a simple way for straight-line geometry. In Fig. 5.21 such a double logarithmic paper is given with the power function geometry on it.

The scale on a logarithmic paper can be obtained in two steps. The first one is to give 10 to each equal 6 sub-lengths on both axes with 9 unequal sub-lengths inside. The second step is that depending on the data value each 10 is given an integer power. For instance, in Fig. 5.21 the power started from 0 up to 6. On this paper both axes are divided into equal distances all with 6 cycles. In the same figure according to Eq. (5.18) there are two straight-line geometry in the forms of directly and inversely proportionalities. For both straight-lines, the slope is numerically equal to n but with opposite signs. In order to understand what the scale parameter a means on this new coordinate system, it is sufficient to consider $\log y = 1$, which corresponds to $y = 10^0$, and hence, the intercept on the vertical axis of 10^2 is the value of a. This can be further interpreted as $a + n \log x = 0$, and therefore, one can obtain $x = e^{(-a/n)}$.

In actual plotting situation with data at hand, in order to decide the power of the least value position on both axes, it is necessary to express the minimum values in

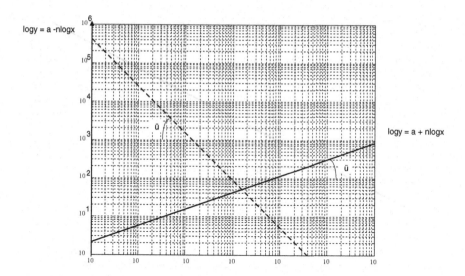

Fig. 5.21 Double logarithmic axes

terms of 10 to the power of some integer. Once the minimum axis position power is decided then the others can be increased as the sequence of real numbers until the maximum value power is reached. Of course, in practice, it is not possible to have exact power value that corresponds to the minimum or maximum value; therefore, the next smallest (biggest) powers are taken for scatter domain of the points. Suppose that the minimum (maximum) value of x data be 0.012 (2457.8) and y values with 27.9 (379.8). In order to plot the points on the logarithmic axes, these numbers must be converted to equivalent 10 to the integer power values. The corresponding values are 1.2×10^{-2} (2.4578×10^3) for x and 2.79×10^1 (3.798×10^2) for y, respectively.

Example: Plot the following x and y values in Table 5.2 on similar double logarithmic paper.

Table 5.2 x and y data

x	y	x	y
1	3.2	1×10^0	3.2×10^0
3	8.2944	3×10^0	8.2944×10^0
5	11.06998	5×10^0	11.06998×10^1
7	12.7713	7×10^0	1.27713×10^1
9	14.6277	9×10^0	1.46277×10^1
11	20.3283	1.1×10^1	2.03283×10^1
13	24.9082	1.3×10^1	2.49082×10^1
15	23.3904	1.5×10^1	2.33904×10^1
17	26.7909	1.7×10^1	2.67909×10^1
19	29.1216	1.9×10^1	2.91216×10^1
21	30.3917	2.1×10^1	3.03917×10^1
23	31.6082	2.3×10^1	3.16082×10^1

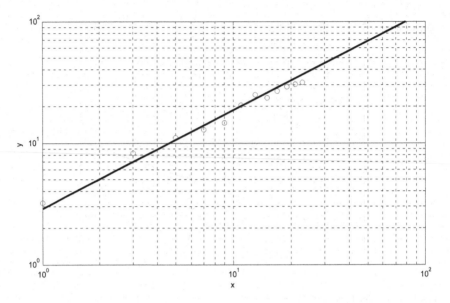

Fig. 5.22 Plotting and scattering on double logarithmic paper

The scatter of points are obtained as in Fig. 5.22, and they all lie around a straight-line, which implies that the mathematical formulation that represents the data best, or the relationship between x and y is a power function.

Exponential curve. Another frequently confronted curve in practical engineering studies is increasing or decreasing exponential function. It can be described linguistically as the curve that intercepts one of the axes and has a single asymptotic value. Its general expression takes the following form.

$$y = ae^{-bx} \tag{5.21}$$

Herein, a is again a scale parameter, but b will be referred to as the branching parameter. Linguistically, it is a continuous function that has an intercept on any axes and has infinite value as an asymptote to another straight line on the Cartesian domain; it can take various geometrical (functional) shapes as in Fig. 5.23.

The family of curves in Fig. 5.23 is representatives from Eq. (5.21), but the ones in Fig. 5.23 result from the following similar expression with the change in the sign.

$$x = ae^{-by} \tag{5.22}$$

Linearization of the exponential curves is similar to the case of the power function but on a single axis logarithmic paper, which is referred to as the semi-logarithmic paper. By taking the logarithms of both sides with respect to the base e, one can obtain,

$$\log x = \log a - by \tag{5.23}$$

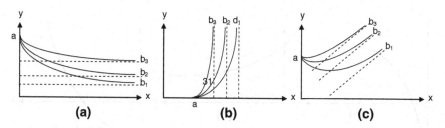

Fig. 5.23 Exponential curves

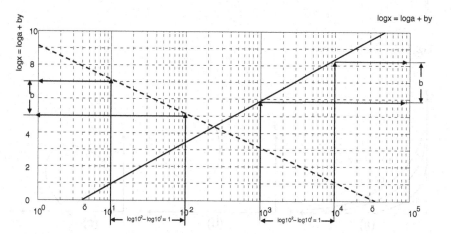

Fig. 5.24 Semi-logarithmic paper

Herein, u = logy and S = loga, and therefore,

$$u = S - by \tag{5.24}$$

Again comparison of this expression with the linear straight-line expression in Eq. (5.2) yields to another straight-line equation. Accordingly one can deduce linguistically relevant interpretations and consequent information. The geometrical forms of Eqs. (5.21) and (5.22) are given in Fig. 5.24.

In this figure d indicates the slope of straight-line on a semi-logarithmic paper. Since, the slope by definition is the ratio of the opposite to adjacent side, on the semi-logarithmic paper, it is better for slope calculation to take the adjacent side length equal to one cycle as indicated in Fig. 5.24. Whichever cycle is considered its length is equal to 1 in natural domain, i.e., without logarithm. In this way the denominator in the slope (b) definition remains equal to 1, and hence, the difference corresponding to the one cycle length yields the slope directly as in the figure. For instance, let us consider the cycle $\log 10^2 - \log 10^1 = 1$ or similarly $\log 10^5 - \log 10^4 = 1$, or any other cycle, one obtains always 1. These are shown in Fig. 5.24. The vertical

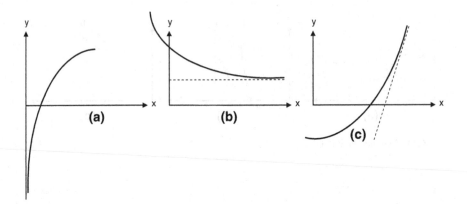

Fig. 5.25 Logarithmic curve

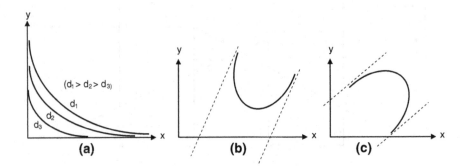

Fig. 5.26 Double asymptotic curve

differences that correspond to each one of these complete cycles yield the same value, b. On the other hand, from Eq. (5.21) for $y = 0$ one can find that $\log x = \log a$ or $x = a$.

Logarithmic curve. Such curves can be identified from the visual and linguistically available knowledge as any curve that has asymptote to one of the axes, intercept on another axis perpendicular to the first one and then increases without any asymptote into the infinity for both axes values. This is a logarithmic curve of which the mathematical expression can be written as follows.

$$y = a + b \log x \qquad (5.25)$$

The most significant difference of this curve from the exponential ones is that it does not pass through the origin (or zero value). The linguistic information about the curve yields various alternative traces as in Fig. 5.25.

Hyperbola curve. Verbally any curve that has two asymptotes along two axes is a hyperbola. These axes may be perpendicular or not. If the two axes are perpendicular

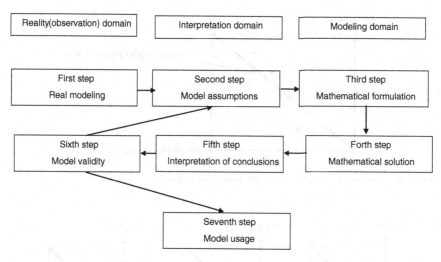

Fig. 5.27 Modeling stages

to each other the simplest form of a hyperbolic curve has the following mathematical expression with its various traces (functions, geometries) in Fig. 5.26.

$$y = \frac{d}{x} \qquad (5.26)$$

In this expression d is referred to herein as the distribution parameter. The smaller (bigger) is this parameter the curve that is asymptotic to two axes gets bigger (smaller).

5.10 Mathematic Modeling Steps

The reliable in any modeling should have the three interrelated branches as in Fig. 5.27. These three paths are reality, interpretation and modeling outputs. The interrelationships between these three paths are shown by arrows [8].

The purpose of modeling is to predict the behavior of the phenomenon or event based on part information, records, measurements and observations. Theoretically, it is necessary to confirm the validity of the prediction results with comparison to the observed data at hand. The first advice for such a comparison numerically is the use of the relative error concept. If the model outputs are denoted by O_m and the measurements as M_m, then the relative error, α, is defined as a percentage according to the following expression.

$$\alpha = 100 \frac{O_m - M_m}{O_m} \qquad (5.27)$$

In practical applications, the numerical value of the relative error should be less than ± 5 or ± 10 % for the acceptance of the model performance. If there are

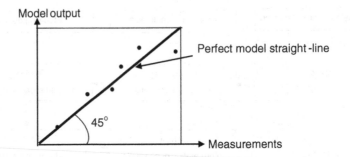

Fig. 5.28 Measurement-prediction (model output scatterplot)

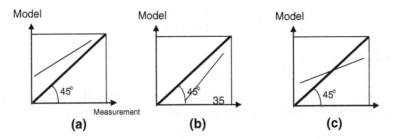

Fig. 5.29 Measurement-model discrepancy

many measurements and corresponding model values then each one of the relative error amount must be less than these percentages. However, some researchers may take the arithmetic average of the whole relative errors and then look for whether it is less than the acceptable percentages.

Far better control for the validity of the model is through visual inspection, which can then be calculated numerically. For this purpose, the sequences of prediction and measurement values are plotted on a Cartesian coordinate system as in Fig. 5.28, where the model predictions (outputs) are shown on the vertical axis and the measurements on the horizontal axis. In case of 100 % conformance between the model outputs and measurements, all the scatter points fall on the 1:1 (45°) line, which is not desired case, because this means that the model predicts the phenomenon 100 %. Such a model is perfect, but it cannot be acceptable in practical studies. Any model will have certain errors, which must be on within the acceptable percentage ranges.

In case that the model cannot represent the phenomenon in a good manner, then one of the alternatives appears as in Fig. 5.29. For instance, over-estimation (Fig. 5.29a), under-estimation (Fig. 5.29b), partial over- and under-estimations (Fig. 5.29c, d), no matching (Fig. 5.29e, f) situations are among the most encountered cases in practice.

References

1. Adler A (1958) The practice and theory of individual psychology. Littlefield, Adams, Patterson
2. Benjamin JR, Cornell CA (1970) Probability, statistics and decisions for civil engineers. McGraw-Hill, New York
3. Felton S, Buhr N, Northey M (1994) Factors influencing the business student's choice of a career in chartered accountancy. Issues Acc Educ (Spring):131–141
4. Freud S (1894) The defense neuro-psychoses. In Rieff P (ed) Early psychoanalytic writings. Collier Books (1963), New York, pp 67–82
5. Kuhn TS (1996) The structure of scientific revolutions, 3rd edn. University of Chicago Press, Chicago
6. Popper KR (1957) The Poverty of historicism. Routledge and Kegan Paul, London
7. Saliba G (2007) Islamic science and the making of the European renaissance. MIT Press, Cambridge, ISBN 0-262-19557-7
8. Şen Z (2002) Bilimsel Düşünce ve Mühendislikte Matematik Modelleme İlkel, Su Vakfı
9. Şen Z (2011) Fuzzy philosophy of science and education. Turkish J Fuzzy Syst 2(2):77–98
10. Von Karman T (1911) Nachr. Ges. Wissenschaft. Göttingen Math. Phys. Klasse pp 509–517 (1911) and pp 547–556 (1912)

Chapter 6
Education and Engineering

6.1 General

Today specifically four-year bachelor engineering education systems are almost stagnant with memorable, traditional, imitative, repetitive and unquestionable information transmissions without interactive debate and criticism. Engineers try to find solutions either as ready procedures, algorithms, formulations or in modern times through internet facilities and software. Even though it has many traditional aspects, static education system and modern engineering topics did not lose much significance along the development path. In engineering education institutions rather than information and scientific knowledge generation, more practical static and unproductive methods, algorithms and software are in use, which may kill the ambitious willingness for information generation. In this manner, the art side of engineering has become weaker and weaker. Computer programs and software rendered engineering education into a stagnant routine sequence of automatic procedures and automation molds without mental, rational and logical inferences.

On the other hand, engineering education system exterminates philosophical thinking principles, and hence, this career is far from artistic and innovative aspects in many educational institutions all over the world. Non-existence of basic philosophical principles in the curriculum also hinders logical linguistic inferences, and hence, engineering takes the form of repetitive steps without many inductive or deductive inferences. It is a sad situation to see the engineering education system as routine career training without much creative activities. Engineering design relies on the following philosophical points, which are missing in today's educational systems.

Z. Şen, *Philosophical, Logical and Scientific Perspectives in Engineering*,
Intelligent Systems Reference Library 143, DOI: 10.1007/978-3-319-01742-6_6,
© Springer International Publishing Switzerland 2014

- Geometry is the most significant topic for engineering and architectural design, which has been indicated as the most significant aspect by Plato, who has written at the entrance of his academy about 300 BC as,

  whoever does not know geometry cannot approach the academy.

  This indicates the significance of geometry for creative human mind.
- On the other hand, Ibn Khaldun, a Muslim thinker, in the fourteenth century stated that

  any mind equipped with geometry does not err.

  This also supports the idea that geometry is important,
- In engineering, design is the main activity in the forms of various sketches, drawings, plans and it depends on the description of thought system, which includes some philosophy,

  any mind equipped with geometry does not err.

- In general, there is not any human thinking system without philosophical ingredients. Otherwise, critical discussion, suspect, innovation and continuous developments cannot be generalized to enlighten the society. Since, engineering is also related to enlightenment, it must have philosophical ingredients.

Unfortunately, today it is almost impossible to see enough philosophical trace in engineering education. Consequently, any engineering candidate takes the engineering methods, formulations, algorithms and software without proper sense and grasp. Engineering information are given like pills and when they are taken, there is no need for further discussion because such pills are regarded as healing solutions for the problem at hand without any additional work apart from the existing knowledge state and know-how only.

6.2 Methodology

Any engineering training center should also lean on the mental scientific contributions on global scale. Unfortunately, many engineering educational centers have concentrated on the applied scientific works, which had imitations and non-creative achievements until very recently. Graduates of these institutions have spread all over the world for the solution of engineering problems and they have shown great success throughout the years for the application of what they have learned during the university training. Scientific questioning is very few, if any. So, it is necessary to establish a new pattern of training, which should take into consideration not only the present day knowledge and information flux to young minds statically, but also the notion of science philosophy and logic with all aspects including research methodology. One of the preliminary prerequisite for such an orientation is the concepts such as science history, philosophy, sociology and physiology aspects of far and near pasts. In this way, those who have ambitions for further research activities will know what the previous researchers in the history have done and what have been their difficulties. This will teach the students of engineering the following significant points

- Ambition and orientation are not sufficient for scientific achievements but also patience and endurance are necessary prerequisites;
- Scientific thought is spatio-temporally continuous process and runs after the criticism of the present knowledge level so as to develop better knowledge sources;
- Scientists and not academicians are among the individuals with different social, administrational and physiological difficulties.

Another simple methodology for better training in any education system is to consider questioning and suspicion opportunities about any type of knowledge. In fact, if the knowledge is not rational, logical and systematic, which means to say that it is not in accordance with the previous knowledge sources, and not grasp-able without memorization, then any individual with research spirit, including engineer should not accept the reservation of such a knowledge in his/her memory. Every scrap of information must be filtered through rational, logical and system-atic learning media so that it remains comfortably in the mind and when the right time arrives its transfer to other individuals becomes quite easy with productive results. The main keys for a dynamic knowledge reservation are suspicion through internal questioning and to ask for explanation from concerned instructors (teach-ers, professors). For dynamic knowledge preservation and generation, leading to new scientific achievements, it is necessary to have a two-end process. These two ends are occupied by the instructor as teacher, on one hand, and student as learner on the other. If the dialog of knowledge flux between these two ends does not function properly then the community is loaded with knowledge that is apparently useful but actually unproductive because it does not open ways for innovative gen-erations. An instructor who allows questioning with rational and scientific answers helps to develop and spread knowledge towards further knowledge generation in the community. Otherwise, the instructor is a murderer for not the body but the mind and soul of the learner. One can imagine how an individual behaves when his/her mind is killed or rendered into non-functional forms. Education is not to teach what somebody does not know, but it is to teach how the mental state should react for solutions in problematic situations. One can ask what the benefits are of questionings in a class-room. Questioning unfolds the following four advantages in order.

- The one who asks the question will satisfy his/her quest for knowing in a sound, clean, proper and systematic way.
- The second benefit is for those who feel rather shy to ask questions, but have queries in their minds. Hence, they benefit from the questions of others by hear-ing when the instructor gives answers.
- Question means criticism, fuzzy understanding, non-systematic conception, etc., which can be clarified only with the answer of learned and scientifically

minded instructors. In this way, everybody in the class will have the right answer to the question asked and when they move to other places after the dismissal, for instance back to home, they will convey the scientific news to others, and hence, in an indirect way, the teacher will be benefactor for the society at a larger scale.

- Perhaps the most significant benefit from questioning is the renewal of the teacher's knowledge. S/he may not have come across with such a question before, and either the answer is given instantaneously in the class based on the correct knowledge of the teacher, or if s/he cannot answer s/he may carry this question to his/her office, and prepare himself/herself to answer properly in the next lecture. Hence, the teacher enlarges his/her domain of knowledge. This is a very good example for two-end process whereby the teacher gives orientations to the students, and the students render the teacher to think and ponder continuously. Such a dual intercourse is very necessary for the generation of new ideas, opinions and knowledge, which might lead to further scientific and technological achievements. Especially, in engineering education, such a two-end process functioning by questions has utmost significance, because any miss-knowledge, -information or -idea might lead to disastrous consequences that may cause loss of material and especially human life.

Furthermore, the last benefit is very useful for the dialog and knowledge flux between student and teacher, who is knowledge fountain that does not stop generating new information. Now, one may ponder about these benefits and then evaluate the education system in any engineering institution in the world. It is better to emphasize at this point that sound, correct, generative and useful knowledge do not come through reading only, but more effectively by listening and discussing. However, in either case the basic key for learning is the criticism and questioning.

One of the very famous philosophers of science [11] explains scientific criticism, thinking, answer producing and development very nicely by saying that:

> We start, I say, with a problem, a difficulty … At best we have only a vague idea about what our problem really consists of. How, then, can we produce an adequate solution? Obviously, we cannot. We must first get better acquainted with the problem. But how? My answer is very simple: by producing an inadequate solution, and by criticizing it.

Science history and especially philosophy become very important at the postgraduate and doctorate levels. In English the title Ph.D. means "Doctor of Philosophy". This is tantamount to saying that anybody with Ph.D. degree should know at least in his/her topic of doctorate the philosophy of the subject that s/he is concerned with. Philosophy means not mathematical symbols, formulations or computer algorithms but their linguistic explanations.

Science cannot be abandoned in engineering institutions for the sake of practicality and applications. Not the engineering subjects only but at early classes the science history and philosophy must be explained at least along flood lines or between the sentences, so that the students will know what they are dealing with and how science and technological developments took place throughout the centuries. Otherwise, graduates including post graduates will not know the answer to the question of "what the science is."

6.3 Education Process

In general, engineering graduate training takes four years and it has been empowered in recent years with courses including social contents, but unfortunately, not yet with courses towards philosophy of engineering, critical reasoning and logical rule inferences. Yet philosophical principles such as integrated wholeness, verifiability and truth, reality and ontology, critical reasoning has not entered engineering training at sufficient levels. Philosophical principles are necessary in any career and especially in engineering, which tries to improve civilization ingredients. At least during each course philosophical thinking and logical principles must be given partially for triggering minds towards inventions. Additionally, each course must be given in an integrative manner with others. If there is not sufficient domain for questioning and philosophy, then thinking cannot flourish towards productive ends. Especially, during education system, perceptions of visual and linguistic knowledge without criticism and doubt cause unproductive approaches in engineering. Let along idea generation, ready formulations and equations in front of engineering candidate become as prescriptions for pills without critical and creative reasoning. Any knowledge set may have then its template that is expected for use at any time and place of need without elasticity in thinking and creative intelligence.

In any engineering education system to empower graduates in an effective way with criticizing ability, course contents must include reasoning in an integrated manner so that engineering career can reach wholeness. During education, the candidates must be taken out for field works, practical trainings, library and laboratory works and computer usages. In such an education system, engineering candidates will feel more knowledgeable at the end of each semester and at the time of graduation they will have self-confidence, self-reliance and independence in thinking and ambitions to make inventions and innovations.

Today engineering has many branches and disciplines, among these the most modern ones are mechatronics, industrial engineering, space, genetic, electronics and many other engineering areas. Each one of these should have philosophy of engineering principles to be more productive in future. Unfortunately, in any engineering education system most often numbers, equations and symbols are dominant in static forms. Engineers frequently start their career without sufficient verbal and philosophical principles.

Does the university education provide information about know-how process? Many staff members can answer to this question in an affirmative manner. However, such an affirmation is not valid absolutely. Philosophy can be interpreted distinctively from know-how as knowing the knowledge in dynamic, explicit and meaningful manners. The process of knowing should remain behind the university education at primary and especially secondary school levels. Universities are education establishments, where knowledge must be processed through various reasoning means (philosophy, logic, rationality, etc.) for additional knowledge generations. If only static knowing processes are dealt with then the society will not advance and become stagnant. One must not forget that enlightenment and productive education should go through philosophical process.

Another question is whether engineering education is for carrying a diploma with static, unproductive and memorable knowledge (algorithms, formulations, equations, methodologies) in a systematic manner? Or is it a training, which empowers engineering candidate with personality, dynamic knowledge and consciousness about what s/he knows? During education, one must try to prepare his/her own road map under the light of the courses in the curriculum. Such a goal can be achieved not with relationships that fall outside the science, but rather with scientific knowledge that are useful in practical engineering applications. In the preparation of road map, each individual must question the initial plan and one must seek also others (teachers, staff member, experienced engineers, etc.) for help. Any thought with others' share requires philosophical basis. One must design the road map not with static boundaries, but dynamic roads with questionable, suspicious and fruitful thoughts.

In many classical education systems engineer candidate indulges with unquestionable knowledge intakes without philosophical, logical or rational thinking and such candidates cannot develop themselves due to inputs of a static education system. Many staff members are also aware of such a stagnant education system and they complain about the system and some of them in spite of the static system try to give students critical and rational reasoning principles. Anybody who enters engineering education system has certain abilities, which must be sharpened during the education, but static inputs cannot provide such a mind sharpening procedure. However, philosophy of engineering principles provides initial thought appearance in a dynamic manner in the mind and then such empowered engineers try to be creative for new ideas. Without philosophical knowledge, information will be inscribed on the minds statically. Such inscriptions cannot provide colorful solutions in front even almost similar problems.

Another aspect of engineering education in addition to rational thinking and ability is to generate the final application products with sense organ (hand, eyes) abilities. In short, an application work can best be achieved by "observation-experimentation-theory" triple with continuous questioning as initial and boundary conditions. The primary foundation of this triple function is the philosophical thinking. Observations provide information by visual appearances of objects, taste, hearing, touching and smelling senses and their rational processes. First appearance helps engineer to imagine the object in mind through perceptions, which provide a basis for different logical inferences and knowledge productions. Hence, these preliminary thoughts and information trigger other rational and mental contemplation in the mind and tell one what to do next. For example, setting up some propositions leads to rational inferences and to their verifications through experiments. For additional knowledge, experiments are necessary but even during the experimentations critical reasoning must continue, which means to say that all the time philosophical and logical affairs must remain at service. After rational and experimental knowledge comes the time for theory and formulations, where it is useful to cite in modern education system that: *Whatever is heard is bound to be forgotten; visualization of things remains in mind; but done things can be understandable*, as the basic criterion in learning and teaching. The basis of education system should not be memorable knowledge in the life road map, but it is better to store information and knowledge with criticisms that will help dynamic visualization in the memory.

The main purpose of education is to transform minds in a selective, open, transparent and dynamic form equipped with knowledge in a variety of ways for further knowledge generation. To know "knowledge", for the sake of perception only, implies its "storage" in the memory without any process. Real education should teach one how self-thinking is? In engineering, getting rid of static and dogmatic practical knowledge are possible through the principles of the philosophy of engineering. It does not mean that any knowledge at the end of any thinking process will be correct. Critical questioning suggests the degree of falseness and how to reduce towards to a better form, satisfactorily accurate information and practical end products. In engineering, instead of accurate solutions, simple, approximate and rational solutions are the main goals of improvement. If such an education system is not available then engineers will remain at static and non-productive levels. In such a situation, engineer may request more information for trying to handle productive conclusions, solutions (analysis) and completions (synthesis). If engineer does not question his/her career problems s/he ends up with nonproductive decisions.

6.4 Engineering Education and Philosophy

In order to obtain better results from an engineering training, instead of repetition, memorization and crisp perception of knowledge, education system should reflect critical discussion, logical and rational thinking. Criticisms must not be limited; otherwise philosophical thinking may be driven away from the education giving way to memorization. Philosophical and logical bases, in arriving to known conclusions, throw away the limitation boundaries, which mean that any knowledge can be criticized continuously. Philosophy of engineering principles can lead engineers to innovative inventions, findings and even to innovative technologies. Invention is related to the accumulation of knowledge in the memory with criticism and rational mind activation near the boundary between known and unknown (uncertain, metaphysical) worlds. This indicates that for creative and innovative findings one should engage his/her mind not within the known domain, but comparatively more near the unknown domain. If the engineer remains within the known domain, then s/he can repeat the same information without any search for betterment or generation of new ideas, and hence, becomes an engineer without the philosophy of engineering. Interest, excitement and ambition for research are all fueled by philosophy of science and engineering in particular, and basic philosophy, in general.

If in universities and research institutions the right for explanation (authority) remains with the instructors, then such units cannot produce knowledge, science and technology. Dynamism can be achieved by giving at least equal chance to the attendants as students. In fact, in any education system clients are students, who request information in return of payment (either by government or private enterprise), and therefore, clients' requests must be satisfied with guidance of sellers (i.e., instructors). Such a liberal education system cannot be established if philosophical principles are missing. Philosophy arena provides each share holding the same opportunity for questioning and answering with criticisms. The most important duty of an instructor is to stimulate the students for criticism until they understand the explanations linguistically in a rational manner. In an engineering education system, linguistic information can be converted later into symbolic and mathematical abstract expressions. Without linguistic bases abstract mathematical expressions are hung in the air only leading to blind applications and unique standard solutions. Answers to any question must not be individual (subjective) but a common mind (objective) production. For such an opportunity, it is necessary to descent down to the philosophical level and leaving academic and career titles aside, debates must take place on equal footing level, and the mutual benefits must be rational inferences from logical and engineering philosophical propositions. Putting the titles aside does not mean that the two sides are on the same knowledge level, but it provides spiritual comfort in debates on both sides. On free thinking horizons, there will be a flow from potential knowledge level towards lower side, but this does not mean that potential side remains on the stage all the time; the balance may shift towards lower side and vice versa.

In many education systems and also in engineering, do training mechanisms, that gives rise to inscriptions of static knowledge gathering; empower the attendants with independent, creative and innovative inventions? In order to provide a continuous and dynamic knowledge production, is it necessary to have only engineering aspects or social aspects, human philosophy or any other science? Are there linguistic information and knowledge generations based on philosophical fundamental and principles that lead to logical rules, and finally, produce dynamic ends? If any engineer sticks to the last part of the sentence, s/he must start to train himself/herself with philosophical and logical principles with linguistic information accumulation rather than symbolic or numerical solutions directly. It is obvious from the history of science that since the first human beings astronomy, meteorology, physics, chemistry and similar natural sciences have started within the philosophical thinking circles with accumulation of new knowledge through many years, where there were not mathematical or scientific principles. However, they achieved all such knowledge and their applications for the benefit of human beings by using their minds, memories, philosophical and logical principles in a dynamic and progressive manner. Is it possible that information and knowledge about the natural phenomena can have continuity and dynamism without inert human abilities such as mind, memory, rational thinking, doubt, and especially, critical discussion? Continuous, innovative and dynamic knowledge generation cannot be achieved without philosophy and logic. Molded static engineering knowledge cannot provide such a dynamic development. Static, memorizable and dogmatic knowledge and information remain as wild giants, but engineering philosophical thinking can overtake these giants and direct them to the service of men.

It is also necessary to keep balance between scientific wings of engineering with special art. Engineers must renew their knowledge after graduation through courses offered by engineering societies or chambers not only for new information collections on specific engineering topics, but also on their functions based on philosophy of engineering and logic rules. During such trainings not only affirmative and useful behaviors, but also bad and undesirable examples must be provided. In developed countries engineers cannot sign plans or any contract right away after graduation, but they need to pass through a professional examination worked out by experienced engineers after certain periods. Hence, a new graduate must wait for some time in order to gain experience and then s/he is entitled to apply for such a professional examination. Only after the proficiency examination, an engineer is entitled to sign under professional reports, plans and documents. After graduation prior to proficiency examination an engineer may experience ethical and aesthetical aspects of engineering in practical life and s/he should also try to have training from philosophy of engineering point of view. Any engineer should have the following points among his/her classical training aftermath concerning the career.

- Fundamentals of physics and mathematics;
- Social, economic and cultural activities and developments in engineering history;
- Design of engineering projects and after their assessment for the application, engineer should provide abilities and different opinions;
- Benefit from experienced and expert engineers' knowledge, information, view and questionings.

The first three can be obtained during the university education and the last one through, so to say, "Life University." In general, neither university nor post-graduate education provides formal philosophy of engineering [19].

Perhaps one of the main reasons of this is that at each stage there are not rational criticisms. However, providence of such knowledge at the second stage may provide engineering benefits, not only during the graduate study duration, but also after graduation during the career life, leading to more productive consequences. Otherwise, physics, mathematics and social topics cannot be stimulating without philosophy of engineering. Integration and dynamic nature of the philosophy may give the same properties to engineering so as to reach the best desired target. Social and humanitarian topics help engineers to integrate with the society and linguistic information attached them together. Philosophy of engineering helps to reach to a single best and optimum solution among many alternatives in the most rational manner. Without philosophical principles, decisions cannot be sound and likewise engineering formulations, algorithms and equations cannot be documented in the form of computer programming (software), because they need philosophy and logical background structure.

In the past, engineers were almost addicted to repetitions and memorizations, but today they are under the pressure of modern thinkless and mindless usage of software, which extinguish the creative ability and fruitful interpretation without philosophical and logical bases. It is better to have simple solutions with philosophy of engineering principles rather than complicated solutions with software without knowing what goes on internally. All software and mathematical formulations have basic linguistic foundations. If an engineer does not have preliminary foundations of philosophy of engineering then s/he cannot question software or mathematical formulations and cannot produce even improved ones. Apart from philosophy of engineering, the formulations and software are accepted without any criticism and they are used as black-box models that do not yield any information about the generation process of outputs under the effects of a set of inputs (Chap. 4). Without philosophy of engineering any approach remains at technician level, however, even a technician with the principles of philosophy may be more fruitful in knowledge generation than an engineer.

It is necessary that engineering education must pass through the questioning and philosophy of engineering criteria. An engineer can provide service to any society through various engineering activities, structural designs and management procedures, but such services may be dangerous during medium or long-terms. For instance, for the benefit of a company or local administration, withdrawal of groundwater by very strong pumps does not abide by engineering aesthetics and ethics. Today detailed stages of an engineering education system should have the following steps.

- To teach fundamental engineering, mathematical and scientific methodologies and to arrive at rational inferences;
- In cases of necessity, as for preparation of experiment and apparatus designs numerical, and especially, verbal information play dominant role;
- To have ability, methodology and software for processing of numerical and verbal information;
- Try to reach the target for the required needs through useful and systematic designs;
- Definition of engineering problems, their solutions, formulations and application facilities;
- To have principles to abide by career and ethical subjects and feel responsibility for this purpose;
- To exhibit not only local solutions to engineering problems, but also at spatial, regional and global scales and then their integration for wholeness;
- To be conscious that engineering problems cannot be solved only according to university education training, but also experience gained after graduation with criticism and interpretations;
- To complete devoid points in the preliminary information by considering modern approaches and methodologies;
- Try to share engineering knowledge in complementary manner in an inter-career team and to reach at integrated solutions;
- To know the use of practical techniques and modern methods with instrumentation and engineering applications;
- At least one commonly used foreign language is necessary to follow technological and scientific developments all over the world.

If there is a question as to where is the philosophy in the aforementioned steps, since engineering is transfer of scientific and technological thinking production into practical applications, at each step philosophical principles should help to develop active and alert thinking channels. By making use of few or all of these steps, an engineering production can be brought into existence, and hence, their functions are completed, but philosophical thinking is never complete and it is continuous towards more fruitful and better consequences.

After all what has been explained above, it is obvious that engineering cannot be separated from the philosophy completely. For such a situation, engineering and philosophy domains must have common area (Chap. 3). The more is the common area, the more will be the critical development, and hence, there will be an engineering horizon open to innovative and dynamic activities in knowledge generation.

The most important and required points in an engineering education can be summarized as follows.

(1) Determination of engineering problems, synthesis and increasing solution capability;
(2) Increase of engineering solutions that will respond to needs by planning, design and verification;
(3) Development of common work abilities with different engineering disciplines;
(4) Team works especially with those from the same engineering discipline, but with different ideas;
(5) Encouragement of student contributions to various activities in engineering education system in various faculties and departments;
(6) Provide an academic atmosphere that is suitable to students' social activities;
(7) Spread of engineering career understandings and ethical values during engineering education;
(8) It is important to give some principles to students for critical discussion in all works during the education period;
(9) Development of engineering skills in presentations, software and report writings;
(10) To provide fundamentals for engineering problem solutions by daily methodologies and approximations;
(11) To provide information increment abilities for life-long engineering works even after the education period;
(12) Information share with other individuals for the solution of various problems;
(13) Increase of engineering views and abilities for engineering applications through modern methodologies and techniques;
(14) In different design projects at decision making stage try to improve personal connective usage;
(15) Keep high level and ambitions in preparation for engineering career after graduation.

6.5 Engineering Education and Problems

Among the definitions of engineering are also the practical deductions after the cheapest, fastest, easiest and satisfactory solutions and selection of the most convenient alternative for the problem at hand. If one examines this definition closely then some restrictions come into the view. The most important of these is that an engineer uses the end product of scientific works for practical applications. However, it is advised herein that not only the use of scientific end products, but also their practical derivations should be known by the engineer. Otherwise, an engineer appears as a science technician without know-how practice. Many engineering institutions educate engineering candidates on more or less technician level. Engineer should not be a man who knows knowledge without detail and dynamism but s/he should be knowledgeable and knows what the information means and how it can be adopted to practical life problems. It is necessary to reason about the knowledge generation and how it can be used in practice. If one does not know the main reasons of the knowledge that s/he works with, then s/he can be treated as a classical technician, who can solve traditional practical problems easily. However, if the problem is little off line, then s/he cannot find solution except by reasoning and expert views provided that s/he has heeded for such reasoning through many years of experience. The question is,

> does the engineering education, today, bring the candidates to the level of technicians only by classical information transfer?

or

> are they trained during the engineering education as those who are able to question, reason and doubt about the solutions along the search of the best ones?

Even though an engineer may not be able to know detailed theoretical information in his/her domain, s/he can find practical solutions depending on his/her previous experience and basic knowledge. This is the most important difference between an engineer and scientist. An engineer should be able to suggest the most convenient and practical solutions in front of complex and expensive problems. The first step towards this goal is to analyze the problem from different angles and then to reach a preliminary solution, which can be criticized and improved by time. Such a process can achieve at the end the most convenient solution if the engineer cares for logical, rational and alternative solution generations. The sequence of his/her mental thoughts brings complex and rather uncertain problems into the domain of determinism with practically absolute solutions. The final decision must be verified from different points of view and at the end plans and projects must be prepared accordingly.

In the present engineering education system, the training remains on "knowing" level and knowledge process does not work properly. Among the reasons are transformation of knowledge (especially by staff members), uncritical knowledge loading to minds, not to run after innovative procedures, students' acceptance of given knowledge without any question or doubt but as verified information even though innovative approaches are necessary but since they do not bring any benefit, engineers use classical and traditional approaches. Additionally, inexistence of science philosophy and logical reasoning in the engineering curriculums add to mechanical training.

In every education system, there must be logical propositions and inferences for the selection of the best rational solution. Any education without philosophy leads to slavery in thoughts and additional illogical training kills the generative fruitful creative thinking. Inexistence of philosophy implies acceptance of given knowledge without any criticism in a dogmatic manner and does not yield bundle of alternative solutions. Unawareness about the logical rules does not give chance to engineer to select the most rational, convenient, practical economic and swift solution among the alternatives. These lead to memorization, dogmatization and direct acceptance through memorization; in addition to formulation slavery in engineering activities, which may even cause unhappiness and turmoil in the society. Instead of the slogan

knowledge is power,
scientism is power

takes place, which may create unrest, injustice, bribery and respect instead of ability. Although enlightenment is talked frequently, but those who advocate it may hinder its entrance into the society. If philosophy, critical debate, logical rules and principles are not available in an education system then politics and even ideologies start to play role. There is no democracy in science, and hence, in scientific education systems. Is it not possible to drive away political and ideological aspects from an education system, if science philosophy, mutual critical debates and logical principles are absent?

In order to promote civilization in any society, engineering activities must take place without biased views. Almost in any activity of the society one may talk about the engineering aspects. Today in health technology, engineering (bioengineering) helps to develop medicine apparatuses where scientific and technological views and aspects play the most significant role. In this manner, engineers support medical services towards perfection. Scientific researches and their end products provide a wide support for many activities in a society such as communication, transportation, management, agriculture, energy, etc. Recent developments in the area of artificial intelligence and their applications in various activities promote engineering career to higher levels. Today science and technology aid for the comfort and management of various tasks in the society. For such active situation, engineering knowledge and know-how must also remain active.

Another important point that must be thought at this stage is that in many societies university engineering education does not provide anything except formulations, equations, algorithms and software without philosophy and especially logic. How can an engineer then without logical steps and rules write software or solve problems that are not classical? Especially, mechanical software usage kills the ability of an engineer in sensing the fundamentals of the problem solving. It is also another indication that today there is an enormous increase in the number of the so called scientific papers, most of which are mechanical and provide application of ready software without philosophical and logical inferences. It must not be forgotten that at the basis of all knowledge are words, concepts, terminology, logical sentences, propositions and their complex expositions leading to objective inferences as conclusions (Chap. 3). Idea generation factories have as their dynamos philosophy and logic after which quality control leads to the final decision. In the scientific aspects, the quality control can be achieved through the experimentation or application and then measurements and observations so as to decide about the success of engineering tasks and implementations. Philosophy and logic as couples imply not numerical information but linguistic knowledge and reasoning with judgments. Even after approximate reasoning, information is kept in the minds as verbal knowledge, which can be used in many cases for problem solving. At times of need an engineer can employ these verbal information readily for use in solving many similar problems with approximate alternative outlets. However, when the problem solving comes at the stage of numerical calculations then these verbal information can be put into the form of formulations and equations, and hence, given numerical input values, outputs can be calculated numerically, which is one of the main tasks of engineers. Without knowing the problem on verbal basis, the application of available formulations numerically may lead to astray. However, after the philosophical and logical principles and linguistic relationships on rational basis and formulations, the problem can be solved appropriately. Linguistic knowledge may require local and slight changes in the existing formulations. Any engineer, who can visualize knowledge in terms of linguistic expressions rationally, can improve his/her sense of grasp in a dynamic manner leading to update his/her information in an expert view manner. This is not possible for an engineer, who relies solely on ready software, formulations and equations numerically. S/he can be very good in manipulation of equations and symbolic information but creativity lacks as long as s/he ignores philosophical and logical principles. Whenever one mentions about an expert engineer, s/he implies the ones that are capable to express their views first linguistically to others even to those who may not be very specialized in his/ her domain. Engineers who rely upon the ready formulations and equations cannot be generative in their career. It should be emphasized that verbal information and knowledge remain dynamically in an intelligent manner in the mind and they can be used as the time comes for practical applications. Especially, a scientist and to a certain extent an intelligent engineer should be capable to translate linguistic information and knowledge into symbolic logic (mathematics, equations) and vice versa. Sole symbolic descriptions intact of imaginative, descriptive and creative artistic works are confined to steady state intelligence without any new productions or innovations. Prior to mathematical symbolic logic, the very basis of thinking in terms of imaginations must be transformed into descriptive forms, which is possible only through geometry. In Islamic countries, the word engineer is "muhendis", which means "geometry knower". How could an engineer make design without geometry? In fact, during the scientific evolution geometry came before mathematical symbolic expressions many centuries ago before Christ.

An engineer must try to explore himself/herself during the education period and after the graduation from physical, spiritual and intellectual angles. Unfortunately, only physical aspects are given the most significance with ignorance of other aspects, which are also important in any creative work. Integrated personality and healthy career are possible with the satisfaction of all aspects in harmonious proportions, which give way to intellectual comfort and idea generation possibilities. Critical questioning must start during the education stage.

6.6 Fuzzy Education System

Logic helps to distinguish beneficial from unbeneficial knowledge through reasoning with arguments, which are means of idea generations from already available knowledge sources through rational inferences. Logic plays a significant role in the theory of knowledge (epistemology) and creation of additional advanced and improved knowledge. If logic prescriptions for reasoning about a certain phenomenon take place then one cannot walk towards better prescriptions if the science and engineering philosophy is not cared for. Although after each philosophical, logical and rational thinking ready prescriptions for problem solving are essential, but what is not essential is their mechanical grasps without arguments. Logic provides and classifies the structure of propositions and arguments and at the end gives a systematic deduction procedure. Two-valued logic is crisp at its consequences, but daily reasoning has causality, possibility, probability, statistics and fuzziness, i.e., uncertainty.

Fuzziness in thinking is indispensable and better understanding of the problems is possible through such uncertain domains. Uncertainty and especially knowing in fuzziness do not provide certain single solutions but they expose a set of alternative generative solutions. It is the task of an engineer or researcher to choose the best suitable solution to the problem at hand. Knowing things and continuous knowing processes include at every step fuzziness that provide a common domain for individual and collective thinking and reasoning.

Logic searches for the meaningful propositions among many sentences in a text or paragraph. Not all the sentences have logical structure and only logical ones lead to thinking, interrelationship existence among various categories and deduction of a final decision. It is, therefore, necessary to have some guidelines for the identification of logical statements in a given text or to construct them in the thinking process about some phenomenon. The simple way of searching for a logical statement is to find one or more of the following logical words. These are,

- "AND", this is one of the logical conjunction words, which joints two categories or statements in such a way that they both are included in the final decision. In this book this is referred to as "ANDing" operation.
- "OR", is another logical conjunctive word that takes into consideration two categories and leads to a common deduction (decision) such that common parts of these two categories are the constituents of the deduction. It will be used as "ORing" operation in this book.
- "NOT", as another logical connective is the negation of the original category. For instance, if the "engineer" is the name of the category "NOT engineer" includes everybody who is not engineer. This will be referred to as "NOTing" operation.
- IF...(A)....THEN...(C)...., is the proposition that includes the very strict logical statement with useful interrelationship among various categories in the antecedent (A) part with the consequent (C). Antecedent part is concerned with the conditional statement (causative part) between the two words IF ..... THEN.

The consequent part comes after the word THEN, which includes the final decision or deduction. A good statement is one whose conclusion follows from its antecedent part. Such logical propositions are at the fundamentals of any education system crisply or in fuzzy manner. However, in practical life almost all deeds include uncertainty, and hence, fuzziness to a certain extent. Even crisp statements are cleaned out (defuzzification) from the fuzzy counterparts.

Expert views cannot be without fuzzy ingredients in the reasoning procedure. Otherwise if crispness was the only way, then different experts could not have consensus on the same problem. It is the fuzzy content of the arguments that provide dialog among the experts or engineers. Unfortunately, most often the engineering education systems are based on crisp logical statements, and therefore, uncertainty and especially fuzzy content of the arguments are driven out from the curriculums. Crisp education system gives the impression that the conclusions are true, which is not so as already explained in Chap. 2 for the safety factor and its meaning. It is most possible that antecedent part of the logical proposition also may not be true. In the classical engineering education system, it is believed that:

if the antecedence is true, then the conclusion is also true.

This statement can be considered as true provided that a set of assumption in the theory, formulation or algorithm is considered as true. It means that the validity of this statement is true in idealized world but not in real situation. However, additionally it would be better to teach engineers also that whatever is the problem for solution one should keep in mind that:

if the premises are fuzzy (uncertain), then the conclusion is also fuzzy (uncertain).

This last statement gives more freedom in thinking, because there may not be a set of restrictive assumptions or simplifying arguments. The comparison of these two statements indicates that reasoning, questioning and criticism freedom is more in fuzzy domain than crisp case. It is, therefore, advised in this book that the education system (especially engineering education) should be based on fuzzy principles, which may later be defuzzified to enter the crisp formulations, procedures and algorithms, if necessary. Some may argue that fuzzy education system is unnecessary, but in social, economic and political sciences this is not the case. However, especially those who are in the numerical domain such as physicists and engineers may be more against the fuzzy education system; however, since the last four decades fuzzy algorithms, inference systems and arguments have entered into almost all aspects of engineering through the expert systems [13, 17, 18, 21, 22]. Those who are against the fuzzy education system may take the course of probability, statistics, stochastic and chaos methodologies and may think that they are dealing with uncertainty. However, they are mistaken because fuzzy system deals with linguistic uncertainties, not numerical uncertainties. Fuzzy education system gives the individual ability to think between true and false in a grey manner rather than black-white two-valued selections. It also provides a common basis to construct dialogs with those who have not higher education, but have rational reasoning such as inventors and innovation followers in the public.

Traditionally in any education system crisp logic is much easier to reach to a common agreement because at the end it provides truth and precision even though in the idealized world. However, the classical education system renders the students to become acquaintant with precision by saying that science provides precision, which is not acceptable absolutely. Fuzzy education system provides open-mindedness in the argument of any problem and open for further criticisms, which helps to improve the present solution alternative, if any.

In this context, fuzzy education system is control system methodology that lends itself to implementation in systems ranging from simple, fast, economic, and systematic solutions. Fuzzy education system provides a simple way to arrive at a definite conclusion based upon vague, ambiguous, imprecise, noisy, or missing input information, and hence, it also enables one to mimic decisions in a faster manner with consensus by taking the opinions of other experts.

The scientific knowledge cannot be completely verifiable or falsifiable but rather it is always fuzzyfiable which provides potentiality for further researches. The science and any related attribute to it will never be completely verifiably or falsifiable but always fuzzyfiable, and hence, further developments in the form of prescience, traditional science and occasional revolutionary science will be in view for all times, spaces and societies [2].

6.7 Education and Uncertainty

Education is a terminology that is used to enlighten others through a sequence of systematic courses that include basic concepts, which are expected to provide for the students a vivid domain of idea creation by pondering on some phenomena. Of course, in such as training, the rational thinking is the core of creative and free opinion. Besides, education has three main facets that should contribute interactively for fruitful and even emotionally stable end purposes. Figure 6.1 indicates these three ingredients in their interactive courses.

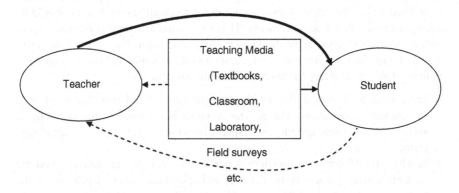

Fig. 6.1 Education system parts

It is obvious from this figure that education does not mean knowledge influx from teacher to student only, but in an effective education system, there are instances when sudden and rather unexpectedly knowledge either flows from the student to the teacher or new knowledge emergence through mutual discussion right during the course. Unfortunately, in many parts of the world and especially in the developing countries, very classical educational systems are in functioning, which provide crisp and elusive knowledge but locally valid certificate in the form of mass production. The defective points in any traditional education system can be specified as follows [19]:

- There is a though authority of teaches, who are directed according to a set of state or traditional rules, which do not give freedom of creative thinking. In such a system, logic means any answer to any question as either black or white. This is classical logical attitude towards the problem solving, and hence, the freedom of the middle categorization has been forbidden.
- Teaching media, which can be referred to as educational gadgets, may become indispensable organs and they are exploited in a crisp and rather dogmatic manner without change throughout years. In fact, in non-native English speaking communities, such devices may easily become show off instruments for affecting the attention of learners to technological wanders rather than basic educational concepts through rational reasoning based on the philosophy and logic.
- There are expectations of ready answers to questions in textbook style of information, which are only jointly shared by different learners and teachers alike without uncertainty or fuzziness.
- Scientific concepts are provided in a crisp manner as if there is only one way of thinking and solving the problems with scientific certainty.
- In crisp logic education system assumptions, hypotheses and idealizations are the common means for mind to grasp the natural phenomena, and therefore, any scientific conclusion or equation is valid under certain circumstances.

In a modern and innovative educational system almost all the concepts must be provided with uncertainty flexibility (fuzzy) especially at the higher educational systems. It is fixed from the long history of science by experience that not only freedom of thinking, but also suspicion from scientific conclusions should be incorporated for better advancements. The very word of suspicion leads to expectation and even viewing scientific knowledge as uncertain. Hence, the basic principles in a modern and innovative educational system should include the following points, which are contrary to classical or traditional education.

- Traditional and classical elements must be minimized and even dismissed from an innovative education system. The authorizable teacher is the one who is authorized as knowledgeable, and especially, who has the ability of knowledge giving.
- Teacher should not be completely dependent on educational gadgets, and the students through discussions, debates and questions should try to force the teacher on the margins of the presented material for more information.

- It should be kept in mind that each scientific conclusion is subject to uncertainty and suspicion, and hence, also to further refinements leading to innovative ideas and modifications.
- Especially, the logical principles and philosophical basis must be kept in the education agenda by the teachers so that each student can grasp and approach the problem with his/her abilities.
- At higher educational level, scientific thinking must be geared towards the falsifiability of the conclusions or theories rather than exactness. Falsifiability implies in itself fuzziness in each decision, theory or proposition.

Provided that all these points are considered collectively, it is possible to conclude that modern and innovative educational training should include philosophical thinking and then logical trimmings, which imply fuzziness in any scientific training. This implies that the conclusions are acceptable with a certain degree of belief (fuzziness) that is not completely certain. The graduates must be confident that there is still domain for themselves to make creative inventions and scientific discoveries in their future. Otherwise, a classical and traditional educational system with the certainty principles in the conclusions do not leave any room for future developments, and consequently, graduates from such systems may hold only the certificate and dogmatic knowledge. However, with the advancement of time in their later ages, they may be frustrated that the knowledge they obtained during their education were not certain, but uncertain, fuzzy and variable for betterment with time.

Traditional or classical systematic education system may give comfort for teachers, but unfortunately it kills the functioning of young minds at the hand of those respected teachers in the society, especially in developing countries. Teachers in such educational systems may become 'mind killers' but they may respected by higher authorities, where the students' requests or representations are not considered at all.

Successful application of the classical control systems is in use for the industrial and engineering solutions. However, there remains still uncertainties to a certain extend that cannot be modeled by the classical approaches, and therefore, uncertainty assessment methodologies are necessary. On this regard, for many years probabilistic, statistical and stochastic approaches and methods have been exploited to the farthest extend by engineers. However, the basic knowledge and information content of the problems could not be appended linguistically to the whole system. On the other hand, their inclusion in the system brings further dimensions and additions, which may be unmanageable to solve with certainty. In this respect, for the last four decades, and especially during the last decade, increasing use of fuzzy systems approach has penetrated many branches of science and technology. The requirement of modern processing industries and technologies encourages flexibility, which results in highly non-linear system behaviors that are known only partially or still there are uncertainties in the main system. These uncertainties may be negligibly small but their collective and especially cumulative effects give rise to complex solutions that cannot be determined uniquely. The classical control systems whatever their advancement levels are can be able to satisfy the demands only partially. However, the fuzzy logic and system analysis can help at this junction to satisfy the demand through qualitative operator and design knowledge for implementations.

In an effective innovation education system for scientifically productive results, the following points' implementations are necessary. It is assumed at this stage that not only the students are considered for the improvement, but more significantly the staff members must be ready to undertake these implementations.

- General behavior of the phenomenon considered must be explained from different points of view on a philosophical level, which indicates the significance of language in the planning and tackling of the problem. This step exposes the significance of language structural and grammatical features in a scientific thinking procedure.
- During the presentation and definition stages of the problem, by all means the students' contribution must be encouraged through various related questions and views. Accordingly, rather than the unique view and the style of the teacher, the topic is rendered to be the common mental property of the student group. It may not be possible to guarantee 100 % agreement between the individuals, but at least a common consensus may be created. Since the students' ability is not at the same level with the staff member, there will remain fuzzy uncertainties at the minds of some students. This is also useful, because it will give further room for discussion among them after the formal classroom sittings.
- The causative effects on the problem must be identified with all possible detail and verbal attachments of variables. Subsequently, the verbal variables must be ordered mentally in the best possible manner according to their significance in the problem at hand. This stage may be considered as dismantling of joint causative effects into individual effects.
- Among the causative effects, a single variable of interest is depicted as the subject of the problem, and hence, there are causative and resultant variables. As a first stage, it is necessary to consider the logical relationships between these variables. These relationships may be very primitive and indicate direct or inverse proportionalities (Chap. 3). Hence, initially there is a list of logical proportionality relationships, which will be further exploited for the refinement of the problem solution.
- Sub-categorization of each variable with at least two, and preferably, three or more classes. This is the stage where the variable names are attached with suitable adjectives. In this manner, each classical variable is rendered into fuzzy variable with various sub-categories.
- Logical propositions including premises among the sub-categories of causative variables are constituted, and subsequently, each one of these premises is attached with sensible, rational and logical consequent parts of the subject variable after rational reasoning. In this manner, the linguistic structure of innovative education is complete by fuzzy logic principles.
- In order to assure the understanding of the students, it is useful to give a common homework and to request the solution of a convenient problem with their individual abilities and linguistic backgrounds.

It is possible to conclude that the innovative training through fuzzy logical ingredients is completely linguistic in character, which gives basic principles of learning and discussing the fuzzy patches from the complete solution. In this manner, information and knowledge are transferred from teacher to student or vice versa. Furthermore, fuzzy logic training system does not include any mathematical formulations or restrictive assumptions. This implies that in the educational systems the mathematical concepts are not the preliminary prerequisites. It should be stated herein that any statement, which insists that the more the mathematics, the better is the research, is mistaken, because the creative education takes place only at institutions where the philosophical discussions and consequent logical regularizations are plenty.

Present educational systems are rather classical with extensive dependence on crisp and blueprint type of information. In many institutions almost spoon fed knowledge and information loadings on fresh brains are experienced without creative or functional productivities. This is perhaps one of the main reasons why in many institutions all over the world, creative and analytical thinking capabilities are not advanced. Of course, it is easy to mention about the quality of students, but more significantly the quality of staff member should also be improved. In developing countries, it is thought most often that the quality control can be improved through students' quality only, which is a defective approach, since highly qualified staff members may lead to improvements in students' quality whereas the reverse is not true. In classical educational systems, more than basic logical propositions, formulations and determinism are mentioned for problem solutions. Especially, in engineering almost each field study is very different from other sites even though there may be some similarities. Therefore, determinism or crisp informational systems cannot be sufficient by itself for the description of phenomena concerned.

Rather than crisp information and solution techniques, as a first step in any innovative education system, fuzzy logic fundamentals must be provided to the students, because it is the natural logic which has been forgotten unfortunately, due to continuous classical logic training in educational institutions. Prior to any equation proposition or verification by data, fuzzy logic concepts may lead to general solution of the problem concerned. In a fuzzy educational training the causes of a phenomenon must be identified as variables and then these variables are considered as sub-categories, which are then combined together through logic propositions to each other.

6.8 Classical Education Systems

Quantitative research methods have a positivistic basis, which restricts wide scopes of related research due to their narrow and limiting assumptions' set as well as specific and measurable data. Hence, the qualitative aspects of research and education are exterminated from consideration and in a way each individual is forced to think according to a set of standards and restrictive regulations. Especially, the formularized nature of quantitative research methods are mathematically supported, and therefore, mostly preferred by engineers and practical appliers. This gives way to a rather hard trend of education, which may cause to memorizing and stock marketing the soft facts in very deterministic and restrictive forms of notations and description forms of unquestionable concepts and statements. Soft research methodologies are basically verbal and linguistic without mathematical symbols, and therefore, attract the notice of many who are not trained in a very systematic manner coupled with mathematical rules, regulations and formulations. The vary basis of qualitative research techniques is the logical foundations, not in the sense of restrictive two-valued black and white (Aristotelian) logic, but rather gray tone weighted logic that has become to be known as the fuzzy logic during the last four decades. The gathering of misgivings about certain aspects of quantitative research methods is deepened by awareness that in some quarters of the conversation, qualitative research, which expressly observes and analyses context-specific phenomena, is considered the more likely prospect to deliver broad, general stable conclusions about education that can be used by practitioners [3, 6, 15]. In the quantitative research approach, there is not embedded certainty of scientific methods, and it removes barriers of scientific dogmas, but dilemma and doubt associated inquiries are sought so as to expand the knowledge domain of inquiry in problem solving. Such an approach is more convenient in educating youngsters with exploring their ambitions and research directions.

The view that quantitative research produces numerical data in contrast to qualitative research producing textual data (words) is one way, albeit an overly simplistic means, of differentiating between quantitative and non-quantitative research [10].

Quantitative methods in educational research are fundamentally aligned with the successful philosophical-scientific-social legacy of the theologian-philosopher René Descartes (1596–1650) involving,

- A quest for certainty;
- A clear delineation between subject and object;
- A view of progress that is always forward moving toward a united system of knowledge.

Acceptance of these positivist features has facilitated consequently a gradual ascendancy or 'privileging' of quantitative over qualitative methods during the past 300 years [8].

6.9 Total Quality Control and Education Technology in Engineering

Quality is a measure of something to be better or worse than ever and its measure is appreciated by the response of the clients. In education institutions, the quality control measure must be established by the common contributions of teachers, learners, administrators and even the permanent service men. On the other hand, total quality management system involves complete studies for the purpose of achieving better standards than today in the functioning of the whole education system with the contributions of everybody involved in this system.

Education technology is important to understand the science of the system and tools. Here, scientific means that one is rigorous about how s/he gathers and uses data. It is better to have a good multiple measurement system that retains and delights costumers than a highly scientific survey that drives customers away.

Total quality control and management are possible only when an institutional unit is ready from its higher level administrators to the costumers with all intermediate stages for corporation, mutual consultation and total improvements. It must be kept in mind that the education media and the costumers as students will always find the truth. It is globally known, recently, that if an institution is not assessed for its ultimate goal achievements, it cannot be managed properly in an efficient way. Quality measurements are possible by multi-directional and education surveys, which are windows into the rigor of any university or research center.

Teaching context may exert a considerable influence at any given moment. Time pressures, heavy assessment, a "cold" classroom climate, and expository teaching encourage surface learning in all students, while teaching to emphasize meaningful learning, assessing for higher order cognitive outcomes, and a context that encourages attributions of ownership and self-efficacy and learner activity rather than passivity, encourages deep learning [1].

Herein, a detailed account is presented concerning the education technology elements as the purpose, design of learning, evaluation and improvements. The ways of combination of these elements are presented on pedagogical basis. The possible application of an efficient educational technology is suggested for Istanbul Technical University.

The very word "technology" in "education technology" must not be confused with the electronic gadgetry or materialistic facilities only. Education technology is as wide as education itself. It is concerned with the design and evaluation of curricula and learning experiences and with the problem of implementing and renovating them. Essentially, it is a rationally problem solving approach to education, a way of thinking skeptically and systematically about learning and teaching [14]. This means that today although there are every facility in the form of technology and their existence in a teaching institution does not sufficiently imply that the educational training is up to date, but these technologies are preliminary necessities for a better educational training. The reverse is also true and perhaps more effective that developing the curriculum with the contribution of all parts is a part of educational technology and quality control, which leads to sound basis for management of further generative education with new ideas and knowledge.

Engineering education is concerned with the generation of technologies, but it needs educational technology for efficient creative work. Among the engineering education productions of electronic gadgetry are expected, but educational technology must not to be confused with such instrumental consequences. Education technology is as wide as education itself. Its main purpose is to design curriculum programs, their evaluations and by feedback to renovate them. Problems in the education system can be solved through rational thinking approach with thinking skeptically (uncertainty, fuzziness) and systematically about learning and teaching. In such a technology, self-learning is an important gradient provided that collective teaching is innovative, creative, actual problem oriented and ultimate economic value. At best one may have vague ideas about what the problem is and then it is not possible to expect a sound or initially approximate solution until a first acquaintance with the problem is reached. The proposed initial solutions may not be adequate but by criticizing it leads progressively to better results. The four phases of educational technology are setting the purpose, design of learning, evaluation, and finally, improvement. Especially, improvement affects not only the design of learning, but also purposes and evaluation. Education is not only to teach people to know what they do not know, but also to behave as they do not behave. Up to date curriculum development also constitutes a very significant part of the educational technology. For the achievement of educational technology goals, it is necessary to promote the total educational environment and the harmonious and successful interactions of all individuals and things (instruments, laboratories, computer networks, internet, libraries, etc.) within it.

6.10 Basic Education Processes

There are two major procedures in teaching-learning process, namely, associations and cognitive approaches. The former is more traditional and developed mostly with the contributions of educators, psychologists and researchers. They take into consideration that learning is a change in either verbal or non-verbal behaviors, which occur as a result of repetition. The basic idea in such learning is student's experience in repeated associations between a stimulus and the correct response to it. In such an education technology for learning in a particular area, the subject matter must be subdivided into a series of small units and then each unit in itself is divided into small pieces of information; through repetition or reinforcement students learn the correct answers to questions about the small pieces of information. As Skinner [20] suggests, learning is largely the result of accumulation of responses to various stimuli that have been arranged in some logical order. Lectures present factual information related to specific steps of a unit, they are supplemented by reading from textbooks or journals.

The performance of students can be evaluated through frequent quizzes on lecture material and reading assignments. The quiz or test are arranged on the multiple-choice questions, quickly corrected and then returned to students for revision. Unfortunately, in many universities returning exam papers to students is not a common practice, and in fact, return of quiz, test or examination papers to students is regarded as illegal. However, how could the quality be improved without the critics of the students, who play the role of clients in an educational affair? It is the duty of instructor to provide discussion platforms for the students, their questions, and the teacher determines students' grasp of information and readiness to deal with new materials. Additionally, laboratory works provide opportunities for reinforcing students' responses to specific information by placing the information in a work case that requires physical participation.

On the other hand, another teaching and learning method is the cognitive approach, which views learning as a purposeful, goal-directed activity involving the gaining or changing of knowledge, skills, or abilities. Learning takes place by reactions of individuals to organizing and then going beyond the information immediately related to a problem situation in order to develop new meanings for them. As suggested by Hilgard [7] learning is a problematical situation resolving, in achieving this resolution, fundamental relationships, principles or methods are understood, and therefore, made usable by learners. In this approach, the learners are confronted with problems directly and try to find solutions with their previous knowledge levels. In such a procedure, the learners try to develop their problem solving skills with their existing knowledge sources and they may find that the previous knowledge is not sufficient for the solution, and subsequently, seek additional knowledge until the solution of problem is completed or carried to a better level. Students use inductive thinking for problem solutions. Through the inductive reasoning, new principles or meanings are discovered and become usable in the problem solution. Hence, an environment is created for the problem solving process; the teacher is a facilitator in such a process by assigning learners through the process of induction. It is obvious that in this training approach, the role of instructor is very important and both learners and teacher react jointly for the solution of the problem. In this process, there is always knowledge and information flux from teacher to student. Likewise, questions, critics, discussions and comments are common means to elaborate such a dynamic teaching procedure. In the cognitive teaching the teacher documents the problematic situations that remain unsolved in the topic, and hence, directs the students to problems and stimulates their attraction to problem solving with their ability, competence and knowledge. Of course, supplemental knowledge and information are provided by teacher at the time of necessity as a result of student requirement. This type of teaching has highly individualistic nature and each individual confronted with the problem performs his/her ability for the solution. In this teaching procedure lectures or large group presentations are minimized.

6.11 Education Methodology

In general, there are four methodological stages in any teaching-learning process, which are complementary in a sequential manner. These stages are quite independent from each other.

- Planning Stage: This corresponds to the identification of goals in teaching and also personal and material resources should be considered for assisting students in achieving these goals. This stage forms the basis for all other teaching activities. During this stage goals and objectives are stated, subject contents are selected and sequenced, resources and constraints are identified, instructional time is allocated, instructional methods and materials are selected, an evaluation plan is developed, and finally, student characteristics are assessed. These activities in the planning stage vary according to the teachers' purpose, time and subject content,

- Development Stage: The necessary allocation and preparation of personal and material for teachings are arranged in such a manner that they are optimum at the present day context. This stage entails gathering various resources identified as essential for managing teaching activities. Among the activities of this stage are arrangements as scheduling, learning experience sequencing, preparation of instructional materials, and lecture preparations, laboratories and other necessary training facilities. These entail activities with a curricular focus, those that take place prior to instruction and present an extension of planning activities, and activities with an instructional focus, those that take place during or as a result of instructions,

- Management Stage: The transactions between the learner and teacher should be arranged in such a manner that the end product is useful for both and for the generation of new knowledge and information. This stage includes also the execution of teaching activities in a harmonized manner. Planning and development stages' activities are brought together in an effective teaching-learning transaction. Management activities are lectures, seminars, discussions, laboratory conductions, arrangements for personalized instructions, and finally, teaching skill refinements. The management activities encompasses those teaching activities that bring together planning and development efforts as teachers transact with students,

- Evaluation Stage: The end product of educational training should be evaluated for the measurement of effectiveness of the total system. The significance of this stage lies in providing the means for determining the effectiveness of the process, its products and for bringing the process once again. Assessment of students' learning and teaching-learning processes are among the activities of this stage. This last stage of teaching, evaluation provides the means for determining the effectiveness of the previous stages. Such an evaluation must not be viewed as an activity that occurs only once at the end of a course of study, but as a dynamic process that can provide continuous feedback to teachers and students.

6.12 Fuzzy Logic Education

In general, engineering, geological, social, and economic and medicine sciences rather than numbers, qualitative descriptions are dominant at initial information in any reconnaissance study with descriptive linguistic explanations. Ordinary people without proper education think in a fuzzy manner because they do not have proper terminology or concrete scientific laws for the descriptions and modeling of the phenomenon concerned. This indicates the effectiveness and natural features of fuzzy logic, which is linguistic in content, but connective between different categories at the background. In order to distinguish between the classical Aristotelian and fuzzy logics let us consider the statement that 'force is directly proportional with acceleration', which is the principle of the Newton's second law provided that the mass is constant. Such a proposal gives a global logical relationship between two variables, which implies that as the acceleration increases, force also increases (Chap. 2). It is not possible to clearly identify from this statement the following points,

- whether the increase in linear or nonlinear;
- validity domain of both variables;
- what are the sub-domains of each variable?

In any research, these are significant questions that need proper answers. In the classical scientific educational systems, these points can be identified objectively by measurements and observations. However, herein the very word of observation must be closely examined and its meaning must be explained linguistically. Measurements need instruments suitable for the study. However, observations may be achieved by human senses and put into words accordingly. Observations are especially significant sources of information in many branches of science. It is rather impossible for an engineer to set forward logical statements about the event of concern features prior to making effective library surveys or field trips. For instance, in engineering geology domain, each area has its special and different features that are not repeated in any other area completely. Hence, right at the beginning, it is known that the geological set up of different regions will have common specifications, features, trends, descriptions, etc., but even so, there will be dissimilar features also. It is these dissimilarities that make the comparison or deduction of information from initial fuzzy behaviors. This is tantamount to say that naturally geological patterns at different sites are dissimilar to a certain degree of content. For instance, globally two different sites of igneous rocks might have the same rock types, say, granite, diorite and gabbro, but it is not possible to insist that each rock type has the same degree of membership in these sites. From the classical logic point of view, these two sites are identical to each other without any further detailed specifications. However, the geologist will not be convinced at all fully that they are identical, because whatever the circumstances, there are uncertainties linguistically which are fuzzy in content. It is possible to ask what the hardness of the granite is in different sites. In general, they will have hardness but not at equal degree, and hence, the variation in the hardness can be sub-categorized relatively as "low", "medium", or "high" which allows the entrance of the fuzzy concepts into the assessments. This leads to the general rule that in any logical assessment, sub-categories are significant, and it is possible to deduce that the more (the finer) the categories, the better is the description.

In fuzzy logic, the fundamental significance is not the sub-categorization, but the relationships among them. One can summarize that for fuzzy investigation of any phenomenon the following steps are a priori necessities.

- Identify the variables for the description of the phenomenon, such as the "force", "acceleration", and "hardness" as mentioned above;
- Sub-categorization of the variables are adjectives such as "low", "medium", "high", "warm", "more", etc.;
- State proposals between the sub-categorization of at least two variables, which must include the logical connections in sentence forms similar to propositions.

In general, geologists are not very familiar or rather do not prefer to apply mathematical rules in their preliminary field and office works, and therefore, most of the information are in the forms of rather vague statements. This is the main reason why especially in geological sciences the fuzzy logic ruler is preferable. It is possible to state that in every walk of daily life individuals without conscious-ness use fuzzy concepts, but this book gives a formal forum for the fuzzy logic ingredients into an innovative education system.

Fuzzy logic approach provides a way of identifying vague relationships between different variables (sub-categories) that play role in the causal of a cer-tain phenomenon. In fact, the mathematical equations either through analytical or statistical or probabilistic approach might lead to such relations, but most often in engineering they are attached with numbers, where non-numerical effects can-not be taken into consideration. Let us think about the causative variables on engineering groundwater storage in permeable layers. The following is the list of causatives,

- type of rock, whether igneous, metamorphic or sedimentary;
- void percentage in the whole volume;
- fracture degree and clusters.

It is possible to relate each one of these variables to the groundwater storage property and fuzzy pair wise logical statements might appear as follows:

(1) According to rock types

IF rock type is "igneous"	THEN storage is "negligible",
IF rock type is "metamorphic"	THEN storage is "moderate",
IF rock type is "sedimentary"	THEN storage is "high".

(2) On the void ratio basis which is terminologically referred to as the porosity,

IF porosity is "low"	THEN storage is "small",
IF porosity is "medium"	THEN storage is "moderate",
IF porosity is "high"	THEN storage is "high".

It is to be noticed, herein, that the porosity and storage have been divided into three sub-categories each with the specified adjectives.

(3) Considering weakly, moderately and highly fractured rocks, the fuzzy logical propositions can be stated as,

IF fracture degree is "weak"	THEN storage is "small",
IF fracture degree is "moderate"	THEN storage is "moderate",
IF fracture degree is "high"	THEN storage is "high".

Each fuzzy proposition can be thought of antecedent (between IF and THEN) and consequent (after THEN) parts. These pair-wise fuzzy logical statements can be generalized into triple-wise, quadruple-wise, etc. propositions with care. For instance, for the problem at hand, the antecedent part has three variables, namely, rock type, porosity and fracture. Since each variable has been categorized into three sub-categories, there will be $3 \times 3 \times 3 = 27$ different combinations of these sub-categories and each one will be attached with one of the most convenient storage sub-categories. The following is the list of exhaustive logical propositions for igneous rocks only.

Rock type	Porosity	Fracture	Storage
Igneous	"low"	"weak"	"small"
Igneous	"low"	"moderate"	"moderate"
Igneous	"low"	"high"	"moderate"
Igneous	"medium"	"weak"	"small"
Igneous	"medium"	"moderate"	"moderate"
Igneous	"medium"	"high"	"medium"
Igneous	"high"	"weak"	"moderate"
Igneous	"high"	"moderate"	"high"
Igneous	"high"	"high"	"high"

Similar tables can be constituted for metamorphic and sedimentary rocks. Logically, filling the consequent part in this table needs careful reasoning. Hence, a detailed exposition of the problem solution appears without any numerical assessment, which is valid in any circumstance. It is stressed, herein, that in any innovative educational system, a systematic approach must be given to the students, so that they can tackle any problem with logical solutions prior to any quantitative (Numerical) calculations. Besides, correct logical statements will provide the students to appreciate quantification, after the availability of numerical data. In classical education systems, just the opposite direction is adopted, in that the students are given already deterministic equations or algorithms without logical steps, and hence, they are becoming formula, equation, procedure, algorithm, etc., addicted. In their future career, whenever they are confronted with a different problem than what they have learnt in the classrooms, they will still expect ready answers on the basis of previous crisp information. Had it been that they are trained with logical thinking and self-confident logical and rational solutions, they would be eager to attack any problem even the ones that are not directly in their domain of specification, but still in their personal interest.

Another comparison of classical and fuzzy logic propositions can be effectively observed and grasped on the basis of Cartesian coordinate system display.

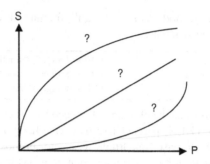

Fig. 6.2 Classical logic domain

Consideration of the above example tells classically that the groundwater storage, S, will increase with porosity, P, which can be shown as in Fig. 6.2. Formally, the classical proposition tells that

  IF porosity increases THEN storage increases

and there are no adjectives in this statement. Non-existence of logical connectives between sub-categories is the main difference from fuzzy propositions. Any crisp logic proposition does not tell whether the relationship is linear or nonlinear.

On the other hand, in the case of fuzzy logic categorization, since storage and porosity are specified by three adjectives, the two axes on the Cartesian coordinate system should be considered as three sub-divisions, which are shown in Fig. 6.3

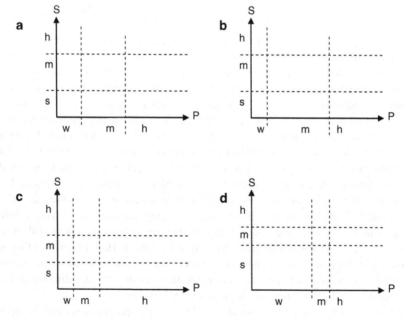

Fig. 6.3 Fuzzy logic domains

for different sets of each adjective. Herein, h, m, s and w letters imply "high", "medium", "small", and "weak", respectively.

Comparison of these four domains with Fig. 6.2 indicates that after the fuzzy partitioning, there are now nine sub-domains each corresponding to a specific relationship between the two variable sub-categorizations. Hence, more detailed and logical interpretations can be done with ease. There are three sources of information for the identification of valid sub-domains for the problem at hand. These are,

- logical deductions, which may be complete work of a non-specialist in the subject;
- expert deductions with specific knowledge on the problem from the previous similar or the same studies;
- data deductions provided that there are measurements or records from previous similar problems.

In many classical educational systems, the students are trained with concentration on the third point whereas the first and second steps are grossly overlooked. Many techniques are thought concerning the third point, especially in engineering sciences through the scatter diagrams and consequent curve fitting procedure by the well-known least squares technique without noticing that this technique has many restrictive assumptions [16].

In innovative educational systems, perhaps, the last point must be left to students more than the two first steps, which constitutes the fundamentals of creative reasoning. Logical deductions should furnish the basis for tackling any problem. If the history of science is reviewed properly, it is possible to see that most of the famous scientists became successful outside of their basic trainings. This indicates that, classical and systematic educational training renders the thinking capability of students into molds with definite boundaries. For instance, if asked about the Newton's law, the ready answer will appear as $F = ma$, or force equals to the multiplication of mass by acceleration. Such minds cannot be creative but robotic, because given the two of the variables (F, m and a) the student will then be able to calculate the third one numerically. Such an approach is a nonsense and nuisance for the prosperity of scientific atmosphere. This is exactly what the third step is in the above explanations. Rather than the formulation, if someone states the Newton's law as saying that the force is directly proportional with acceleration, then s/he has dependence on crisp logical principles to a certain extent. The same saying can be put into a formal form as "IF acceleration increases THEN force increases." This statement does not tell anything about the rate of increase.

Let us consider the same law from the fuzzy logic point of view by sub-categorizing the force and acceleration into three categories as "low", "moderate" and "high". Consequently, there will be 9 sub-domains as in Fig. 6.3 but not all of these sub-domains will be valid logically. Rational and logical reasoning without any expertise or data exposition will invalidate 6 of these sub-domains which are, ("low" acceleration–"moderate" force), ("low" acceleration–"high" force), ("moderate" acceleration–"low" force), ("moderate" acceleration–"high" force), ("high"

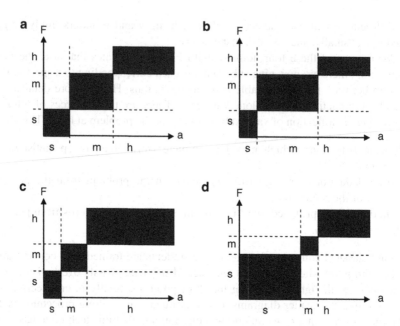

Fig. 6.4 Acceleration-force sub-domains (fuzzy logic)

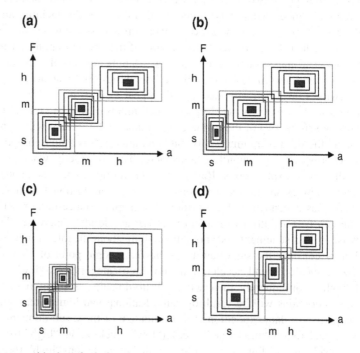

Fig. 6.5 Fuzzy partitioning of sub-domains

acceleration–"low" force), and finally, ("high" acceleration–"moderate" force). Hence, three logical statements remain, namely, ("low" acceleration–"low" force), ("moderate" acceleration–"moderate" force), and ("high" acceleration–"high" force). Furthermore, the graphical representations of these three sub-domains are shown in Fig. 6.4 with different proportions of each sub-domain.

In this figure, black and white sub-domains are still indicators of classical logic remnants, which can be further fuzzified for complete description of fuzzy logic assessments. One of the fuzzy logic properties is that there must not be sharp, i.e., crisp boundaries between sub-domains. This brings still another question as "adjective domains" and what are representatives of "adjectives." The logical answer is that any adjective domain has degrees of representativeness of the adjective. The most representative point in each sub-domain is in the centroid of the area. Such an approach leads to the gray areas in the sub-domains depending on the degree of membership of the adjective attachment to the sub-domain. Hence, the blackness fades away towards the edges and even there appears overlapping between the adjacent sub-domains as shown in Fig. 6.5.

6.13 Fuzzy Philosophy Bases of Science and Education

Engineers cannot be completely objective in their justifications for scientific demarcation or progress but ingredients of fuzziness are driving engine for the generation of new approaches. All logical rule bases must be justified, i.e., tested by physical generation mechanism of the phenomenon under consideration, which lead to scientific domain but for classical understanding and dissemination of the knowledge, the people also need logical verbal explanations rather than symbolic or numerical equations. The scientific phenomena are all uncertain and fuzzy in nature, and especially, the foundations of scientific philosophy include embedded fuzzy components. Dogmatic nature of scientific knowledge or belief, in science as if it is not suspectable, is the fruits of formal classical Aristotelian logic whereas fuzzy logic holds the scientific arena vivid and fruitful for future scientific plantations and knowledge generations. Scientific consequences are dependent on premises that are logical proportions of the phenomena concerned. These proportions are verbal and linguistic statements, and therefore, at the initial philosophical thinking stages they include vagueness and imprecision. As more scientific evidence becomes available, either rationally or empirically, these statements' validity degree increases, or vagueness proportion decreases. In the philosophy of science, so far scientific statements are either assumed as absolutely correct, but more often than this, they are accepted with some probability degrees. However, objective probability attachment to scientific statements is a difficult task, and therefore, subjective (Bayesian) proportions are attached to these statements in practice. After a detailed account of what were the advocators and opponents to scientific absolute correctness and probability, a fuzzy thinking, and consequently, membership degree attachments rather than probability are presented by considering fuzzy subsets in the following sequel.

The term philosophy has a wide meaning including from a cloudy speculative fancy to a piece of formal logic (Chap. 3). Until recently, the formal logic in philosophy has been taken as the Aristotelian logic, which has alternatives of two completely mutually exclusive classes as true or false; positive or negative; black or white; beautiful or ugly; etc. All scientific hypothesis, theories and ideas have been measured first on the basis of this logic, and consequently, classical scientists have emerged at times with dogmatic believes and insignificant scientific developments. The reason for such a majority of academicians who cannot have attributes of sciences or scientists is due to the crisp nature of the classical logic, where even the cloudy, vague, uncertain, imprecise qualities must be crisply classified into distinctive and mutually exclusive parts. None of the scientific knowledge can be accounted as completely crisp without doubt, otherwise scientific developments cannot prosper. The existence of scientific development is not due to the exactness of the knowledge but rather due to its vague characteristics [22]. The use of fuzzy logic is suggested for the demarcation of scientific knowledge and development of science. It is declared that whatever is scientific, it includes fuzzy information to a certain extent and Aristotelian logic cannot be valid in natural environment and human thought extensively.

Common sense dictates that some form of empiricism is essential to make sense of the world. In quantitative educational research, the classical dualism of educational research, the tension between subjectivity and objectivity is often addressed by adopting an objectivist, empiricist or positivistic approach, and then by applying a scientific research design. Scientific thinking starts in an entirely subjective medium of the scientist, with the imagination and visualization, subjective ideas penetrate objectivity domain by time, and hence, there is not a clear cut crisp line between these two media. Empirical works, which are based on either observations or measurements, i.e., experimental information help to decrease the degree of subjectivity. In a way, none of the scientific formulations obtained up to now is completely crisp, but they are regarded as crisp information provided that the fundamental assumptions are taken into consideration. The crispness of any scientific information can be shacked by modifying one of the basic assumptions. This implies that all the scientific principles are not crisp 100 %, but include vagueness, incompleteness and uncertainty even at small scales, and hence they are fuzzy by nature.

In order to move understanding towards a deeper and broader grasp of complexity, the emergent meanings need to be stable but stable enough to rely upon them when generating hypotheses, concepts, and emotional attitudes not to allow these concepts and attitudes to harden and become dogmas and addictions. In other words, meanings need to be flexible, ready for immediate responses to continuous changes in every dimensions of reality.

6.14 Fuzziology

Human beings are created to think and take decision for their daily life activities towards prosperity. They are even referred to as "clever animals", which can judge the circumstances and reach the purpose whatever it may be. Five sense organs provide intake information from the surrounding environment, and accordingly, the decisions are taken after the logical and rational treatments and judgments. However, since the origin of life for many millions of years, the judgments are internally processes by human mind and the results are produced accordingly.

Our mind is the generator of fuzzy impressions and conceptions. It divides the seeable environmental reality into fragments and categories, which are fundamental ingredients in classification, analysis and deduction of conclusions after the labeling each fragment with a "word" such as a name, noun or adjective (Chap. 2). The initial labeling by words is without any motion and interrelation between various categories. These words have very little to do with the wholeness of reality—a wholeness to which all of us belong inseparably. Hence, common words help to imagine the same or very similar objects in our minds. Furthermore, the real world that has numerous fragments can be captured by sense organs' perceptions that trigger thoughts and meaningful inferences. They serve collectively to provide partial, and therefore, distorted conceptual models of reality, which represent a perceived and human-mind-produced world. It is not a world whose natural evolution has brought us to existence and with which human beings are linked through an umbilical cord of vital and impossible-to-separate connections. The basic postulate of fuzziology is simple,

Our understanding and knowing grow from within us and cannot be implanted or imposed from without.

Human understanding and knowing are self-organizing processes; and any self-organizing process in nature works from inside out [4].

Individual awareness is an essential sensor for strengthening consciousness so that any individual can take independent decisions without blind directions. It is necessary to consider the following essentials, if one would like to be alerted along this line.

- Keep in mind that always knowledge come to beneficial forms from unknown world such as metaphysics and once the information is confirmed with philosophical and especially logical rules then it is possible to convince others always or for some time until they give their responses in a debatable forms, which creates an interactive atmosphere for view exchange and ripening the common knowledge and information. It must not be forgotten that creativity, innovation, invention and modifying existing knowledge depend on the existence of uncertainty,
- In education it is advised that anything learnt from parents, teachers and even most reliable thinkers one must not clink to absolute definiteness, certainty and commonsense knowledge products. One must be suspicious and doubtful about what s/he learns until it become very obvious and clear in the stilling basin of mind,
- Try to understand even the most complex phenomenon by breaking its structure into sensible fragments with a set of assumptions, idealizations and isolations and after the grasp of these fragmentations at a certain mind level through linguistic discussions and logical rule establishments then keep the general rule in mind and if necessary try to formulize its functions in mathematical equations for engineering uses. Once the generation mechanism of the knowledge bundle

is understood about the complex or even ordinary events then one can advise its use to humanity (engineers or specialists) so as to fire further intellectual capacity of human beings. In all these achievements the role of spirit must be put into functioning as appropriate as possible.

The limits of present day knowledge can be transgressed by thinking in the uncertain (fuzzy) world for search of systematic generation mechanisms, which keep minds alive, productive and active. Advancement in human knowledge cannot be without uncertainty (especially linguistic uncertainty, i.e., fuzzy) and in this manner human can fight with complexity. The more deterministic are the consequences the more the slavery of mind tends towards dogmatism in science, memorization and prescription applications without any creativeness, innovation and invention. For a productive mind fuzziness in human knowledge is an indispensable ingredient, which gives way to better understanding, assessment, evaluation and inference. A simple rule that can be put forward is that one must not make his/her mind slave of crisp knowledge and s/he should know that precision in knowledge leads to erroneous directions and conclusions. Social, economic, political and even engineering phenomena cannot be isolated from each other completely, and therefore, they always have fuzzy contents, which provide ways for further growth, research and development. Precise knowledge can be regarded as valid in partial solutions after a set of assumptions that should be satisfied at least approximately under the circumstances of the problem of concern. Since we are away from the absolute truth in scientific, philosophical and logical propositions we are concerned with relative truths, which are not intact (mutually exclusive) from the real truth. Hence, uncertainty in the form of fuzziness helps one to search for better truth but not absolute truth by time. Experience (not experimentation), sensing, feeling, understanding, thinking, reasoning, knowing and accordingly acting are among basic human behaviors that cannot be improved in 100 % exactness. These human behaviors are activated by intakes from objects outside the human. Individuals have different consciousness, understanding and inference capabilities and in the fuzzy environment these take their better shapes depending on these circumstances. The more one knows, the more s/he may err, because s/he trusts his knowledge treasure and makes courageous inferences. On the contrary, one whose knowledge is limited, s/he may take certainty for granted without questioning.

In educational system, if the known information is regarded as crisp knowledge, it tends to become a dogma, and the dogma kills ones thinking ability and capacity. For the scientific evolution and advancement, it is necessary that any knowledge should be regarded as fuzzy in the sense that it still includes uncertainties of some kind. Hence, fuzzy education system leaves vital domains in the human thinking for prosperous developments and for generation of various alternative scenarios and then assessment of the optimal (best) solution under a set of actual conditions.

> As the complexity of a system increases, human ability to make precise and relevant (meaningful) statements about its behavior diminishes until a threshold is reached beyond which the precision and the relevance become mutually exclusive characteristics [22].

Fuzzy statements are the only bearers of meaning and relevance. Fuzziness is an essential characteristic of ones imaginations that raise and dissolve in his/her thoughts—in ones memories and reflections about the past and in his/her plans and dreams about the future. They have blurred boundaries and consist of fuzzy immaterial substance. Having in mind how important is to think in images for the development of one's intelligence and capacity to learn and know, to act and create, to evolve and transform, one should not underestimate the role of the fuzziness in human evolution (imagination and description) [4].

Fuzziness is the processes of learning, generating hypotheses and proving theorems. In [5] Gödel proved that in any axiomatic mathematical system (theory), there are fuzzy propositions, which cannot be proved or disproved within the axioms of this system.

Intelligence and wisdom improvements are not easily compatible with the establishment of rigid mental and emotional patterns. Emotions and spiritual internal feelings are among the necessary drives for better knowledge generations. In some countries standard, prejudice and official education systems train individuals blindly, crisply and dogmatically without any constructive debate In such education systems there is not proper directions for students according to their abilities but they are left to make their future according to present fashions, which have mostly materialistic aims. Under such circumstances, students are bound to enter a set of classical careers, and hence, there appear accumulations in these careers. In this manner the capability of youngsters is lost forever. Such mechanisms put sets in front of creative thinking and subsequent productive innovations or scientific researches and advancements.

Present day engineering education system does not provide intensive questioning and critical view to gain experience in the class room from the lecturer. The students memorize the given information and try to solve examination questions during the education in a classical manner and after the graduation in the work imitatively. Such training does not empower the student with dynamic mental abilities and independence. In a way the students are given borrowed knowledge without any criticism, philosophy or logical rules and they cannot even think of altering or modifying current appliances, methodologies or algorithms. The application of knowledge is not thought and what is the benefit of inapplicable knowledge except obtaining titles. Education systems should try and provide applicable knowledge in the field, factory or laboratory during the training period. For this purpose, supportive financial, administrational, and industrial cooperation are necessary. Although there is slogan of state-university-industry cooperation frequently in the media, but in many countries such a consensus has not been achieved yet.

There are many crisp engineering plans according to the client requests. Any engineering structure is subject to natural, industrial and terrorist instabilities, which are uncertain and the designs must be based on uncertain principles. In the meantime, it is necessary to incorporate linguistic knowledge and information also fuzzy logic rules in a reasonable and constructive manner. Expert views and knowledge accumulate through applications by gaining experience in design, implementation, development and continuous innovation of human-made systems.

It is, therefore, necessary that each education system should direct engineers about how to exploit their accumulative information in applications at least practically. On the other hand, artificial human made structures require precision and certainty to utmost level, and this is the trend taken in engineering education institutions all over the world. Engineering is taken as granted for precision, certainty and absolute solutions that are reliable without danger or risk subjects. This is valid especially in electronic engineering but other branches request more uncertainty involvements for problem solutions.

In engineering there are numerical and conceptual models. Especially, conceptual models reflect the verbal knowledge about the generation mechanism of the phenomenon concerned and such models deal partially (in fragments) with the problem solution based on human perception. The fragments together constitute the holistic picture provided that the fragment interrelationships are identified initially by philosophical and then expressed by verbal logical statements.

In modeling the reality, the first approximation goes through the crisp logic where the mind confronts with duality as something and its opposite. This helps engineer to construct the fundamentals of the final goal through dubious feelings about the crispness. The duality is referred to as the binary (two-valued) logic and in this way, the first approximate solutions can be obtained excluding sources of uncertainty. The duality principle brings a restriction on the freedom of natural thinking and logic, because there is no vagueness, ambiguity, possibility or probability. Crisp logic is a sort of trap that channels human mind and thinking into routine solutions, prejudices, restrictions and dogmatic beliefs in science. Although the crisp logic filters human understanding, dynamic minds are suspicious about the consequences, and therefore, may enter the domain of linguistic uncertainty, i.e., fuzziness. It is well known that in daily affairs everybody confronts with vague, ambiguous, uncertain, possible and probable events. These features lead one to enter a wider domain of logical thinking and mental productions. Fuzzy way of thinking hardly excludes others opinion or features of the objects, because almost everything has a degree of belongingness attachment between 0 and 1, inclusive. Such a broad logic does not exclude anyone's opinion about the same phenomenon, but weights it in some way. Hence, human mind dynamism gains activity, because even contra-dictionary opinions, arguments and opinion clashes have some degree of validity, which open the channel for possible consensus among many experts.

After all what have been explained above, one should try and get rid of the restrictive crisp logic domain with fuzzification of all the ideas. This will lead thinking mind towards an arena of intellectual and productive ideas. Anyone who is eager to run after knowing and learning can gain these features by thinking that concepts are in a continuous evolution and better improvements appear day after day. Although there are restrictive boundaries in crisp logic thinking, there is no such thing in fuzzy domain of affairs.

There is a rational belief about the logical inferences, which are not numerical but linguistically changeable as new knowledge and information enter the discussion domain of the problem. However, the fuzzy nature of logical thinking and search for relationship between causative and consequent elements do not cease with certainty. Whatever are the circumstances, there is always some fraction of numerical and verbal uncertainty remnants even in the conclusive final decisions. In the construction of logical rules for explanation of any problem, researchers or engineers should not look for those propositions that can entail, or contradict, but also for propositions that can partially entail, or partially contradict other propositions. According to Keynes [9] probability comprises that

> part of logic which deals with arguments

are rational but not conclusive. The same sentence can be read by replacing probability with possibility, which means to say that premises include fuzzy subsets. Hence, it is a part of logic, but not mathematics.

When evidence changes, vagueness, and therefore, degrees of belief also change in thought rather than in experience, because they are logical relations of partial entailment between propositions expressing conclusions in which one has degrees of belief and propositions expressing the evidence for the conclusions. Probabilities as degree of belief are subjective rather than objective; they represent psychological states [12]. One should not understand the rationality of the probability judgments expressing partial beliefs arising from scientific investigations as a matter, not of their correspondence to something external to them, or of their derivability from a supposedly objective indifference principle, but of the relation of the beliefs to each other.

References

1. Biggs JB, Telfer R (1987) The process of learning. Prentice-Hall, Sydney
2. Carnap R (1987) The confirmation of laws and theories. In: Janet A (ed) Scientific knowledge. Wadsworth Publishing Company, Kourany
3. Denzin NK, Lincoln YS (2001) The discipline and practice of qualitative research. In: Denzin NK, Lincoln YS (eds) Handbook of qualitative research. Sage Publications, London
4. Dimitrov V (2000) Discovering fuzziology. University of Western Sydney, Sydney
5. Gödel K (1932) Ein Spezialfall des Entscheidungsproblem der theoretischen Logik. Ergebnisse eines mathematischen Kolloquiums 2:27–28. Reprinted and translated in (Gödel, 1986, 130–235)
6. Guba EG, Lincoln YS (2001) Competing paradigms in qualitative research. In: Conrad CF, Haworth JG, Latucca LR (eds) Qualitative research in higher education: ASHE reader series, 2nd edn. Pearson Custom Publishing, Boston, pp 57–72
7. Hilgard ER (1996) History of educational psychology. In: Berliner DC, Calfee RC (eds) Handbook of educational psychology. Macmillan, New York
8. Jansen G, Peshkin A (1992) Subjectivity in qualitative research. In: Le Compte MD, Millroy WL, Priessle J (eds) The handbook of qualitative research in education. Academic Press, London
9. Keynes JM (2004) A treatise on probability. Dover Publications, New York
10. Krathwohl DR (1998) Methods of educational and social science research: an integrated approach. Longman, New York, p 742
11. Popper KR (1957) The poverty of historicism. Routledge and Kegan Paul, London
12. Ramsey FP (1978) Truth and probability in foundations. In: Mellor DH (ed) Essays in philosophy, logic, mathematics and economics. Routledge and Kegal Paul, London
13. Ross T (1995) Fuzzy logic with engineering applications. Wiley, New York, p 628

14. Rowntree D (1986) Teaching through self-instruction: a practical handbook for course developers. Kogan Page, London
15. Schwandt TA (2001) Dictionary of qualitative inquiry. Sage Publications, London, p 281
16. Şen Z (2001) Angström equation parameter estimation by unrestricted method. Sol Energy 71:95–1007
17. Şen Z (2010a) Fuzzy logic and system models in water sciences. Turkish Water Foundation, Ankara
18. Şen Z (2010b) Fuzzy logic and hydrological modeling. Taylor and Francis Group, CRC Press, Boca Raton, p 340
19. Şen Z (2011) Fuzzy philosophy of science and education. Turkish J Fuzzy Syst 2(2):77–98
20. Skinner BF (1968) The technology of teaching. Appleton-Century-Crofts, New York
21. Türkşen B (2006) Ontological and epistemological perspective of fuzzy set theory. Elsevier Science & Technology Books, London, p 514
22. Zadeh LA (1973) Fuzzy sets. Inf Control 8(3):338–353

Chapter 7
Future Trends

7.1 General

Engineering career has penetrated into a variety of human activities that it become necessary to look at its very roots for creative reasoning leading to rational and productive products. It has already been noticed since the last decade that this career is away from the basic philosophical reasoning that should be inferred through the logical principles not in the form of final symbolic structures that help for ready engineering calculations, but initially in terms of linguistically propositions. Although the end products of scientific researches in the forms of equations, formulations, algorithms and software are essential products for engineering career, without their logical and verbal expressions, it is not possible to communicate let alone among the same engineering disciplines, but even among the experts, who are capable to grasp linguistic explanations. In order to eliminate such rigid situations and provide a common linguistic discussion domain among all types of expert views, recently science philosophy version of engineering philosophy and its subsequent activity, logic with its propositions started to take place among the engineers towards engineering philosophical aspects. For this purpose, advanced engineering principles based on the philosophical reasoning and logical inferences are bound to play significant role in future engineering activities. Such ingredients are necessary for innovative creative end productions with research and development activities.

The focal point in any engineering activity is the design in a systematic way. If the word "systematic" implies crisp rules and regulations then there is no way that philosophical principles can penetrate into engineering. It is, therefore, necessary that the engineers should adopt rather elastic or fuzzy designs to open door for philosophy towards setting up principles of philosophy and engineering. Additionally, if engineering is taken as science production and mathematics, again, although there are philosophical aspects behind these disciplines, their penetration into the engineering is interrupted or hindered. How can science and mathematics be without philosophy and logic, which is emphasized as the basis

Z. Şen, *Philosophical, Logical and Scientific Perspectives in Engineering*,
Intelligent Systems Reference Library 143, DOI: 10.1007/978-3-319-01742-6_7,
© Springer International Publishing Switzerland 2014

of mathematics? Here comes the difference between traditional mass production engineering education mechanically and future engineering expectations with empowerment of philosophy and logic. It is also possible to state that prior to any mathematical and scientific bases, say before renaissance, engineers were more craft oriented, where they have to think linguistically and had philosophical and logical trends. However, the so called modern engineering trainings and education systems confined engineers within philosophy and logic impermeable domain. The relationship of engineering with science and technology compels it to get involved with philosophy and logic. In another way of expression, science runs after discoveries, technology inventions and innovations whereas engineering searches for the proper application of the scientific and technological end products, but this does not mean that there is not feed-back from engineering to technology and science, presently. Hence, these are intermingled with each other and their genuine bases are philosophy and logic. Furthermore, science develops theories, technology patterns and blueprints and engineers' material products and design are based on the theories and patterns.

7.2 Research Interest

In recent years, engineers started to become interested in logic and systems approach in various aspects of science, which requires verbal knowledge and information, but still there is not enough cover in all engineering aspects. However, the number and variety of such applications are increasing ranging from single element identification to complex system modeling. Natural phenomena have different types and varieties of internal and external factors that give rise to the occurrences of engineering phenomena (earthquakes, droughts, floods, sedimentations, dam designs and failures, landslides, etc.) in a sophisticated manner and such complications cannot be explained completely by classical crisp mathematical formulations. It is, therefore, very efficient to consider the basic philosophical fundamentals and logical foundations of these phenomena in order to reach reasonable conclusions. Although probabilistic, statistical and stochastic approaches are used to model uncertainty for many events to investigate them scientifically, but they all depend on a set of assumptions and besides need data for the model establishment. On the contrary, in daily life and especially during the initial confrontation with engineering events ready formulations, software and algorithms cannot be quit useful. At the initial stages of any problem confrontation mathematics, differential equations or engineering symbolic approaches may lack proper solutions, because they include a set of assumptions and require numerical data for proper and convenient model identification.

However, rational and logical thinking and reasoning have their brain functions that deal with inexact information after the imaginations and fundamental inspirations about the phenomenon concerned. Hence, automatically one starts to consider systems that are dynamic and numerical model free approaches for the solution of the event. Initially there is no need for numerical estimators, because estimations and predictions could have their basic fundamentals within the thinking stages (Chap. 2). Additionally, such genuine and non-mathematical initial thoughts and inferences are also prerequisites of system modeling suits for further research and applications. Such approaches start from highly systematic insights about the phenomenon or event concerned and their categorizations prevail in human mind as for "strong", "weak", "acceptable", "unacceptable", low", "middle", "high", "reliable", "suitable", etc. Uncertain and imprecise information are the basis of the human thinking, which may lead to mental deductions by pondering on the causative and resultant effects. In this manner, complex and non-linear systems can be modeled without the involvement of extensive and rather difficult mathematical abstractions. The basic logical ingredients are in the form of reasoning through simple questions, "why", and "how", which result in IF–THEN rules as a kind of expert knowledge (Chaps. 4, 5, 6). In this manner any engineer or clever men can exploit experience and expertise by lying down if–then statements, which are referred to as the implications in logical context.

Logical assessments require initial mental reasoning for the identification of similarities and differences between various scrap information, and finally, they are established in a systematic way to express the relevant conditions. Such systematic approaches lead to automation and machine intelligence of science phenomena.

Engineers deal with natural phenomena (earthquakes, droughts, floods, space researches, transportations, soil mechanics, etc.) which are full of vague and qualitative information in linguistic forms, as words that imply categorization, shape and size characteristics in the first instance. In general, those who work especially with quantitative assessments need to appreciate the scale of the phenomena or events. Practically, there are four different main scale categories that describe objects in a non-numerical manner with fuzzy implications. These are nominal, ordinal, interval and ratio scales. Each main category may have sub-categories as the necessity requires. It is advised in this book that any researcher should be acquainted with the etymological and epistemological content of each word and its implications in time and space, shape and scale.

7.3 Scale Concepts

Perceptions imply approximate quantities in mind and they can be appreciated and sensed by words for preservation in human memory. Reflections in the memory provide appreciation about scale magnitude together with some meaningful features of the phenomenon or objects concerned. Each word provides bases for appreciation of some qualities in vague and relative quantities for description of objects For instance; a very small piece of rock particle, "silt" cannot be put into "sand" category, provided that one knows etymological and epistemological distinctions between the words, "silt" and "sand". Hence, human mind does

Table 7.1 Nominal scale of "damage"

Without damage	Very minor damage	Minor damage	Moderate damage	Severe damage	Very severe damage	Collapse

not accept "sand" size silts, due to conscious experience perhaps based on logical measures, which attribute a scale to each object through logical comparisons.
Of course, s/he can also appreciate that there is no clear cut boundary between
the two categories, which are silt and sand in this case. For such mental appreciations, assessments and inferences precise measurements are not necessary prior to
problem imagination, description and discussion. An engineer should be capable
to imagine the expected size of result whether it at the scale of a cat, an ant or an
elephant. Perception and the appreciation of scales are significant prior to numerical solutions.

In general, there are four scales that provide the qualitative appreciation of any
object. These are nominal, ordinal, interval and ratio scales. The information content of each one increases from nominal towards ratio scale. During the mind sensing process the qualitative scale appreciations are gained, which is stored in the
memory after the right comparison [1, 2].

7.3.1 Nominal Scale

The word "nominal" implies object categories according to their name attachments. A single word or few of them are sufficient for appreciation of the nominal scale and it helps to categorize a set of objects into almost mutually exclusive
groups. Each category has a specific linguistic allocation that gives appreciation
after comparison with all available categories. In Table 7.1 such a nominal scale
is given concerning the word "damage", which may be the classification of earthquake damage on existing buildings [3].

Here, there is no numerical measure such that any category is, say, twice as
much as any other. Since definitions are made linguistically, computers cannot
appreciate this scale, because there is no numerical value. Crisp logic dichoto-
mizes them into mutual exclusive and exhaustive sets by attaching numerical
boundaries between each category. Computers can then distinguish between
the categories numerically through the two-valued crisp logic, which implies
that the neighboring categories are mutually exclusive. In reality, neighboring
categories are not mutually exclusive, and therefore, in human mind there is
always an overlap between adjacent categories. This point implies that nominal scales have uncertainty, vagueness and they remain as qualitative classifications. Furthermore, it is not possible to quantify or even rank the objects within
each category. They are used frequently by experts for objects' appreciation,
and therefore, the opinion of each expert does not overlap with others 100 %.
Although majority of experts may agree greatly on the same category, but few

Table 7.2 Köppen climate classification system

Climate type (Nominal scale)	Description
A	Tropical moist climates: all months have average temperatures above 18 °C
B	Dry climates: with deficient precipitation during most of the year
C	Moist mid-latitude climates with mild winters
D	Moist mid-latitude climates with cold winters
E	Polar climates: with extremely cold winters and summers

others may disagree and locate the some parts of the same objects into one of the adjacent categories. For example, if out of five bridges subjected to flood event, then only two of them may be appreciated as having "low damage", but it is not possible to rank these two bridges as long as they are within a nominal scale. There are many examples of nominal variables such as climate, morphology, vegetation, land use, etc.

It is possible to convert linguistic information into nominal scales for appreciation of the phenomenon concerned even by non-experts. For instance, the Köppen climate classification system is most widely used for classifying the world's climates. Its categories are based on the annual and monthly averages of temperature and precipitation. The Köppen system provides five major climatic types, where each type is designated by a capital letter as in Table 7.2 [2].

Stevens [4, 5] classified not just simple operations, but also statistical procedures according to the scales for which they are "permissible". A scale that preserves meaning under some class of transformations should be restricted to statistics whose meaning would not change if any transformation is applied to the data. By this reasoning, analyses on nominal data, for example, should be limited to summary statistics such as the number of cases, the mode, and contingency correlation, which require only that the identity of the values be preserved.

7.3.2 Ordinal Scale

The word "ordinal" means ordering the objects according to their ranks. Hence, any rankable data set has ordinal scale. In the ordinal scales, each description is expressed by words that include uncertainty. One can count and order, but not measure ordinal data. For instance, the nominal scale in Table 7.1 can be converted into ordinal scale as in Table 7.3, where aforementioned percentages are attached to each category.

It is not necessary that the interval length in each category should be equal, but in various lengths as in this table. Ordinal scales allow one to order the items in terms of which one has "less" and "more" of the quality represented by the variable, but still they do not allow one to say "how much more." One can say that

Table 7.3 Ordinal scale of "damage"

Group	1	2	3	4	5	6	7
Description	Without damage	Very minor damage	Minor damage	Moderate damage	Severe damage	Very severe damage	Collapse
Percentages	0–10	10–25	25–40	40–60	60–80	80–95	95–100

nominal measurement provides "less" information content than ordinal measurement, but questions such as "how much less" or "how this difference" help to compare the difference between ordinal and interval scales.

7.3.3 Interval Scale

"Interval" as a word implies that there are lower and upper boundaries in each category. This scale allows to arbitrary zero (initial) point value. It is possible to add or subtract scores on this scale, but multiplication and division do not yield meaningful results. For example, Gregorian calendar has the zero (initial) years as the birth of Prophet Jesus Crist, and therefore, we are now in 2013. However, calendar based on Moon, which is used in several Islamic countries is referred to as Higra, which has started with the migration of Prophet Mohammad from Makkah City to Madina City and according to this calendar we are now in 1336. Both of these calendars have interval scales but with different starting points. One can appreciate summation and subtraction operations within the interval scales, but not division and multiplication. Furthermore, temperature degradation is according to interval scale such as (23)–(24 °C) and (1)–(2 °C). The zero temperature represents a natural event of coldness. Hence, zero is attached with a natural event of water freezing in the Celsius interval scale whereas in the Fahrenheit categorization, the zero is attributed to the temperature of snow and salt mixture. This indicates that any interval scale is relative to a natural event.

Each level of measurement includes the measurement principle of the lower level of measurement. For example, the numbers 18 and 19 in an interval scale indicate that the object assigned a 19 has more of the attribute being measured than does the object assigned an eight (ordinal property) and that all persons assigned a 19 have equivalent amounts of the attribute being measured (nominal property). This also implies that one can do lower level statistics on higher level measurement scales. Interval scales do not have the ratio property. Permissible statistics for ordinal scales included these plus the median, percentiles, and ordinal correlations, that is, statistics whose meanings are preserved when monotone transformations are applied to the data.

Interval data allow in addition, means, standard deviations (although not all common statistics computed with standard deviations), and product moment correlations, because the interpretations of these statistics are unchanged when linear transformations are applied to the data.

7.3.4 Ratio Scale

"Ratio" as a word implies division (or multiplication) operation in the calculations. It has a true zero point and equal intervals. This scale helps to convert from one unit to another through a conversion factor. Ratio scales are dominant in mathematics, physics, engineering calculations, probability, and statistic in addition to stochastic, where calculations are performed according to crisp logic. Among conditions of scientific affairs there are time and/or scale dimensions, which are typical examples of ratio scale measurements. For example, the velocity is a ratio scale, not only can one say that a velocity of 150 m/s is higher than 25 m/s, but s/he can add that it is six fold higher.

Finally, ratio data allow to geometric means and coefficients of variation, which are unchanged by rescaling the data. In summarizing this argument [6] said:

> The scale type places [limitations] upon the statistics one may sensibly employ. If the interpretation of a particular statistic or statistical test is altered when admissible scale transformations are applied, then our substantive conclusions will depend on which arbitrary representation we have used in making our calculations. Most scientists, when they understand the problem, feel that they should shun such statistics and rely only upon those that exhibit the appropriate invariance for the scale type at hand. Both the geometric and the arithmetic means are legitimate in this sense for ratio scales (unit arbitrary), only the latter is legitimate for interval scales (unit and zero arbitrary), and neither for ordinal scales.

Any object in engineering can be measured by an instrument and the results appear in numbers. Table 7.4 shows the measurement principles concerning each scale [2].

Table 7.4 The measurement principles

Nominal	Ordinal	Interval	Ratio
People or objects with the same scale value are the same on some attribute	People or objects with a higher scale value have more of some attribute	Intervals between adjacent scale values are equal with respect the attribute being measured	There is a rationale zero point for the scale. Ratios are equivalent, e.g., the ratio of 2–1 is the same as the ratio of 8–4
The values of the scale have no 'numeric' meaning in the way that one usually thinks about numbers	The intervals between adjacent scale values are indeterminate. Scale assignment is by the property of 'greater than,' 'equal to,' or 'less than'	Example: the difference between 8 and 9 is the same as the difference between 76 and 77	

7.4 Future Education Aspects

Many engineering problems are full of qualitative information that varies according to specific situations and in many cases one cannot have quantitative information but qualitative knowledge is available even from non-specialists, because they may have observed various phenomena during their life time. Due to vagueness, imprecision and at places and times incomplete information, uncertainty concepts, principles and systematic approach become very suitable for their application in engineering trainings. Any word terminologically implies linguistic contents in addition to possible time, space, shape and size qualities, each of which plays significant role especially in the early research stages prior to indulgent to any mathematical formulations. Especially, shape (geometry) and size of the objects or events are important for proper visualization and establishment of logical statements and propositions about the phenomena concerned. As already mentioned in the previous section there are four verbal information sources as for the size of an object or event, which are nominal, ordinal, interval and ratio scales.

Traditional or classical systematic education system may give comfort for teachers and students alike, but unfortunately, it kills the functioning minds. The proper and fundamental aspects of a healthy education system is not qualitative at first stages but it should be descriptive, and linguistic leading to logical rules in the form of internal relationships between the causative and resultant variables concerning the phenomena at focus for scientific investigations.

Present educational systems are rather classical with extensive dependence on crisp and blueprint type of information. In many institutions almost spoon fed knowledge and information loadings on fresh brains are experienced without creative or functional productivities. This is perhaps one of the main reasons why in many institutions all over the world, creative and analytical thinking capabilities are not advanced. It is easy to mention about the quality of students, but more significantly the quality of staff member should also be improved. This is achieved in developed countries by yearly contracts, but in many countries the staff members work as government employees. This gives an untouchable status to staff members and whatever they do whether productive, instructive, and innovative or not nobody can move them from their positions. Such life-time automatic status does not give dynamism to the education system. In such situations dynamism is sought among the students only, whereas the total quality of the staff member is assumed as good or the best. In a way node your head and at the end of each month take your salary without any complication.

On the other hand, in classical engineering educational systems, more than basic logical propositions, formulations and determinism are mentioned for the solution of problems. Especially, in engineering and in many other disciplines also almost each field study at a site is completely different from other sites even though they may be close to each other. Therefore, determinism or crisp informational systems cannot provide productive new information for the description of phenomena concerned.

It should be always preferable that rather than crisp information and ready solution techniques, in any adaptive and innovative education system, logic fundamentals must be provided to the students, because it is the natural logic which has been forgotten due to continuous classical two-valued crisp logic training in educational institutions. At many engineering education systems the role of philosophical thinking and logical inferences has been almost forgotten; instead ready formulations, software and algorithms are abundantly given to students. Prior to any equation proposition or verification by data, logical concepts may lead to general solution of the problem concerned. In logic educational training the causes of a phenomenon must be identified as variables and then these variables are considered as sub-categories, which are then combined together through logic propositions to each other.

7.5 Uncertainty Training

Most often engineers are inclined to have solutions that are crisp without any domain for uncertainty. Although there are methods that give room for uncertainty such as recurrence intervals risk calculations, safety regulations, early warning systems, future predictions, but they may not be aware about the techniques used in their assessments, evaluations and calculations. Engineering education system was not effective before 1950s concerning uncertainty, since then although probability, statistics and stochastic methodologies gained steady increase in dealing with engineering solutions, but again their very fundamentals are not known logically and verbally. Many engineers depend upon software for solution of their problems even in the uncertainty domain, but the outputs from such software cannot be interpreted for inferences, deductions and beneficial interpretations and conclusions. Hence, one can suggest that from the early years of engineering training the students should be acquainted with uncertainty principles not through crisp mathematical procedures, but on debatable linguistic knowledge, information and qualitative data bases. The most modern way of treating such data is through fuzzy logic principles.

Since its inception by Zadeh [7], fuzzy logic theory has received wide scope attentions both in its theory development to gain maturity and in its applications including engineering. Recognizing that fuzzy logic is a powerful theory to handle imprecise information, its application has been initiated in decision making process. Recently engineering education systems in many universities all over the world are full of ambiguous, vague, imprecise and random information sources that can be treated with the fuzzy logic concepts for clear ideas and solutions. There are different versions of engineering topics such as deterministic hydrology, seismology, earthquake engineering, hydrogeology, engineering geology, petroleum geology, geochemistry, geophysics, environmental engineering, etc., which require almost certain numerical information for the application of ready formulations or software runs even with methodologies that work through numerical random variables. The engineering educations are mathematically based on the classical logic with two alternatives with the exclusion of middles. Accordingly, mathematical equations, systematic algorithms and formulations are the bases of the modeling for estimation, prediction, model identification or filtering purposes. In probabilistic and stochastic modeling processes a set of assumptions is necessary such as the stationary, homogeneous, ergodic, intrinsic and temporal and spatial independence. These assumptions render complex phenomena to manageable mathematical sizes and domains. Otherwise, the ignorance of the engineer cannot be accounted by classical logic, which constitutes the foundations of mathematical models. The success of mathematical models is dependent on the numerical data base. This is the reason why every research unit in the world would have to have a sound data base. The significant question is what about the logical rule base, which is linguistically data base? Fuzzy logic aspects in engineering incorporate in model construction, processing and control stages non-random (linguistic) uncertainties together with the numerical data base. The fundamental skeleton of such modeling is the rule base rather than data base. Rule base includes all the linguistic data in the form of uncertain (fuzzy) sets with membership functions, which are communicators between the computers and human mentality. Hence, fuzzy logic based engineering can be defined as a new version of engineering research alternatives where a suitable model identification for the problem at hand has logical rules with fuzzy sets as basic ingredients in making relevant dependence statements between the input and output variables of the system. There are two basic types of uncertainty that may be present in any real-world process as follows.

i   Stochastic uncertainty is due to a lack of information, where the future state of the system may not be known completely. It has been handled by probability theory statistics and stochastic processes. The outcome of a stochastic event is either true or false.

ii Fuzziness is vagueness concerning the description of the semantic meaning of the events, phenomena or statements themselves. This uncertainty type implies that fuzzy engineering techniques are necessary for the solution of the problem at hand. In this situation, where event itself is not well defined, the outcome may be given by a quantity other than true (one) or false (zero). That is, the outcome in the presence of uncertainty may be quantified by a degree of belief. The events are modeled as fuzzy sets because the characteristic function of such sets may take the values other than zero or one. Hence, one may talk about the probability of a fuzzy event, i.e. the likelihood that something vaguely defined would happen.

If a real world problem is sought that is inherently messy, where mathematical functions are difficult to apply, uncertainty modeling processes would be an excellent example. Engineering processes, at or near the land surface depend on topography, vegetation, and soil moisture, rainfall patterns and intensity, potential evapo-transpiration, air temperatures, solar radiation, winds, and dew points. Each of the variables changes either in space or in time, and many change in both space and time. Nonetheless, it is necessary to calculate such processes in this real world environment. Engineering models are useful only to the degree that they represent processes in real world.

Although probabilistic, statistical and stochastic approaches and methods are used for many years in engineering, still linguistic knowledge could not be digested. Inclusion of linguistic data in engineering systems brings additional dimension in problem solving. The fuzzy logic and system analysis can help at this junction to process linguistic data. The fuzzy concepts have led to misunderstandings and controversial issues between many researchers.

An engineering design may be defined as that socio-economic activity by which scientific, engineering and behavioral principles together with technical information and experience are applied with skill, imagination, and judgment in the creation of functional economical, aesthetically pleasing, and environmentally acceptable devices, processes, or systems for the benefit of society.

An obvious drawback to fuzzy logic is that it is not always accurate. The results are perceived as a guess, so it may not be as widely trusted as an answer from classical logic. Certainly, though, some chances need to be taken. How else can groundwater engineer succeed in modeling by assuming the average porosity of alluvium deposits is 0.25? Complex interactions among engineering and natural events give rise to the spatial as well as temporal evolutions, which must be controlled in a scientific manner so as to render its consequences beneficial for human activities.

The basic estimation work has been performed by Gauss in early 1800s who tried to fit the most suitable curve through the scatter of points by having the least squares technique as a criterion which constitutes without exception the basis of any uncertainty event assessment in statistics and stochastic processes modeling. The successful application of the least squares technique for almost two centuries is due to the following factors.

i The minimization of sum of squared errors leads to a system of linear equations, which are easy to solve and do not require an extensive theory,

ii The sum of the squares correspond in many different context to various interpretations such as in physics the energy is expressed as the sum of squares; in mechanics it represents moment of inertia, in statistics it provides the variance about the fitted curve and consequently it can be used as a measure of the goodness-of-fit test,

iii An assumption of a definite explicit analytical form to represent the observed data constitutes the principal application of the classical least squares technique,

iv Without proposing an explicit analytical expression, it is possible to apply the least squares technique to filtering problems. For instance, a known differential equation may represent the phenomenon concerned,

v Shannon and Wiener [8] have founded a different application version of the least squares technique by assuming a certain statistical properties for the useful signal and noise constituents of observation sequences. The significant difference of Shannon and Wiener approach lies in the fact that the useful and noise parts are characterized not by analytical forms but by their statistical properties such as the mean values are supposed to be zero or rendered to zero and both serial and cross autocorrelations,

vi After 1950, in order to reduce the computation burden Carlton and Follins [9] suggested the use of adaptive least squares technique. However, Kalman [10] suggested an elegant procedure for the adaptive prediction in the form of recursive filtering. This technique is generally considered as sparked the widespread interest in the subject of estimation, and therefore, it deserves a detailed account. The practicality of Kalman filtering showed its effects in the areas of aerospace engineering and navigation in addition to guidance since 1960.

It is emphasized herein that rather than crisp information and solution techniques, as a first step in any engineering system, fuzzy logic fundamentals must be provided, because it is the natural logic which has been forgotten unfortunately, due to continuous classical logic training. Prior to any equation proposition or verification by data, fuzzy logic concepts may lead to general solution of the problem. In fuzzy logic engineering training the causes of a phenomenon must be identified as variables and then these variables are considered as sub-categories, which are then combined together through logic propositions to each other.

The engineering knowledge cannot be completely verifiable or falsifiable but rather it is always fuzzifiable which provides potentiality for further researches. As a general conclusion of this book, it is assessed that the engineering subjects will not be completely verifiably or falsifiable but always fuzzifiable, and hence, further developments in the form of prescience, traditional science and occasional revolutionary science will be in view for all times, spaces and societies [11].

7.6 Thinking Development in Engineering and its Future

In general, human thinking enables one to establish relationships to subjects in the environment for positivistic inference purposes under the limitations of the societal and cultural virtues. In this manner, internal concentration in the thinking accumulation provides ability to construct rational and logical interrelationships. Although objectivity and subjectivity seem to be mutually exclusive, but their overlapping existence in the engineer's mind may lead him/her to deduce significant conclusions or preliminary inferences about the phenomenon concerned.

Even though objective and subjective thoughts appear as separate from each other, as they go through the mind by time s/he infers some rational deductions and inferences. Although scientific and technological inferences are reached after the positivistic deductions, non-positivistic phenomenal thoughts remain in the thinker's mind. The most important indication of such a situation is after the deduction of someone; others review such deductions and improve them to a better and general scientific or technological level. During the development of scientific patterns, sometimes such deductions are taken as general rules without extreme criticism, and therefore, scientific evolution patters has become rather stagnant with time even though there were not periods of decrease, but some other time periods have occurred with steeper increase in the scientific developments. For instance, Newton's physics have not been criticized for almost two centuries due to its accordance with medium scale event representations. Einstein's relativistic views have shown that it is not valid in large scales such as the light velocity or at very high velocities. In the case of any scientific development there were resistances from others until this scientific theory or hypothesis is verified experimentally. Unfortunately, frequently during the science history different researchers have been against to each other's views, but at the end the most objective and general one has been selected after factual verifications. The more the cooperation between the individuals who are concerned with almost the same problem, the better has been deduction in a shorter time than any single individual. However, individuals also have generated extreme ideas, which have paved ways to revolutionary scientific affairs. In some societies, today, some thinkers cannot come together in a team work, and therefore, the scientific and technological developments appear at the speed of a turtle. In short, complete freedom in thinking is not searched for scientific and technological development, but more than that, it is the improvement of common sense and change of traditional thinking towards this direction.

If different objective or subjective thoughts do not exist in a harmonious manner, then even though extreme thoughts may pave way towards scientific and technological developments, it will have slight improvements. Dependence on the positivistic thoughts only appeared as one of the decelerating effect in the scientific development, because then many subjective information that wait for improvements with previous positivistic knowledge bundles have been driven away from the thinking process. The most significant example for this is the belief in the Newtonian physics until the nineteenth century. According to this belief, it was possible to explain any event in the past or in the future by force, mass, acceleration, momentum and classical energy conservation laws. The following statement was alleged by Simone de Laplace in his publication concerning the first articulation of causal or scientific determinism Laplace (1749–1827).

> We may regard the present state of the universe as the effect of its past and the cause of its future. An intellect which at a certain moment would know all forces that set nature in motion, and all positions of all items of which nature is composed, if this intellect were also vast enough to submit these data to analysis, it would embrace in a single formula the movements of the greatest bodies of the universe and those of the tiniest atom; for such an intellect nothing would be uncertain and the future just like the past would be present before its eyes.

He also stated that if one is able to determine the initial conditions precisely then s/he can predict not only the future but also the past. In this way, the established idea empowers one to calculate time and space places independently from each other. This opinion has been expanded later to cover all sorts of phenomena, and finally, the universe and all the events inside were assumed to work as a clock, which has been set up by a super intelligent existence, God. This idea gave rise to the possibility of controlling any event by mechanical and mathematical formulations. Hence, determinism, homogeneity and isotropy concepts entered into the description of natural and engineering events with simplifying assumptions that led to mathematical formulations. This approach brought a sort of mechanical ingredient into the human thought. As a result of this view, those who thought outside this frame were not looked upon nicely. This is similar to creating idols from the things that have been generated by human, and hence, human began to run after scientific inspirations similar to a religion. However, evolution of scientific and technological works have shown today clearly that such idols have been broken into pieces and in the future even this pattern will advance along the same line.

The first objections against Newtonian mechanics started during the nineteenth century due to its lack in explaining heat problems, which lead to the development of another dynamical branch, thermodynamics. The heat laws have shown that the Newtonian mechanics cannot be valid in any case, and hence, a new scientific revolution emerged. Naturally mechanical and positivistic principles have been shaken from their foundations. The most significant view of thermodynamic approach is to consider the universe as an engine, and hence, as any engine, it is bound to age and loose its performance and function by time towards an end that the engine can no more perform its duties. According to this principle everything in the universe bound to age and become weaker in the performance and at the end reach to a stagnant destination. Such a view has been elaborated even during the old Greek period. For instance, Plato (BC 427–BC 347) stated that the existent objects are grasped by human through their shadows, and in fact, ideas are absolute existences and they do not change by time, but their shadows change temporally. Hence, changes and transformations continue by time and they will also continue in the future. It is, therefore, possible to allege that all the scientific laws of today might change in the future depending on the ageing of the universe. Due to these sustainable changes humans will be involved in the scientific affairs all the time. Parallel to all these advancement engineering processes, affairs and procedures as well as methodologies are also bound to advance towards new horizons.

As a trend from the engineering history, engineers should have philosophical bases in their creative works, but not pure philosophy, instead science philosophy. One should ask at this stage, which type of philosophy should be given during the engineering education today? Is it pure or science philosophy? Since, engineers are concerned with the comfort of the society; in general, leading to objective solutions, science philosophy is necessary in their basic education program so that they can adjust ideas linguistically in a qualitative way prior to numerical solutions. In this way, engineers will be empowered to suggest not only a single solution similar to case studies or problem solutions in text books, but several alternatives, in which case the engineer then tries to select the most rapid, cheap, secure, and optimum solution by reserving other features for future use.

Today's scientific developments have shown that continents were not at their present state and location, but they move in micro scale that cannot be appreciated during human life. This is another proof of evolution of the universe. For instance, about

400 million years ago at the time of joint continents, the weather and climate patterns were different than today. Accordingly, meteorological and climatological facts and laws at that time were different from the present situation. At the moment, the universe is estimated to have 4.5 billion years of age, and its remaining life is about 5 billion years, and hence, today's scientific laws are bound to change in the very distant future. For this reason, scientific principles and their end products used by engineers must be considered as a continuous development and evolution due to present information content missing and ageing.

On the other hand, today science and its product as technology and engineering developments provide new insights and results in human philosophical thinking. Human always prefer "right", "good", "simple", "easy", "beauty" and "homogeneity" with "isotropy". Even at the early stages of grasp many phenomena show nonlinear performances, sophistication (chaotic) and uncertainty. Human tries to sense these as simple as possible. In this way, s/he knows that there are uncertainties in any description of the natural phenomenon, but during the science history frequently employed deterministic rules are used in engineering without any change for arriving at beneficial conclusions. In some societies, determinism has dominated such that they could not be able to consider any uncertain gradient in the scientific and engineering domains. They have labeled those that are in their control with deterministic principles and ignored all others that included uncertainty. Many have had faith in absolute determinism. This has restricted free thinking for further developments in science, technology and engineering. Many societies are not aware of such a hindrance and they have taken the scientific, engineering and technological prescriptions given to them as pills, without any creative thinking. A good example for this is the Euclidian geometry, which has prescribed against the human natural sense the whole geometry in terms of point, line, area and volume in prime dimensions. This geometry has not been criticized for almost two millenniums or even though there were allegations against it, but they were not heeded sincerely, and hence, geometric improvements could not be originated for the service of scientific engineering and technological developments. Einstein could not explain relativistic views by Euclidian geometry and he searched for a convenient geometry and finally found that Riemann geometry is suitable for his thoughts. This is a good example that at times scientific hypothesis and theories become similar to dogmatic beliefs, which do not give way to further developments. It is not possible, for instance, to draw a tree with Euclidian geometry, which has been described recently satisfactorily by fractal geometry [12]. Fractal geometry takes into account not prime dimensions but decimal dimensions also. It gave way to human to generate natural views and figures with ease. Likewise, today instead of Aristotelian logic of 0 and 1 (again prime numbers), fuzzy logic of [13] is based on decimal numbers between 0 and 1, inclusive. Fuzzy logic is a revolutionary scientific principle, which triggered recent technology in many areas bringing into the view the expert views by different individuals in their respective involvements.

Addition of new revolutions to the ones in the geometry and logic domain, as explained above, is possible by avoiding belief and dogmatic principles in the scientific domain. Another restrictive principle in the scientific studies is the concept of linearity almost in any discipline, because all of the laws (Newton, Hooke, Hubble, Ohm, Darcy, Fick, etc.) and especially in engineering are linear in their forms. In short, all laws can be expressed in an English sentence as,

Two variables are directly and linearly related to each other.

This sentence covers all the scientific laws in their linear form provided that the two variable names are explicitly stated. For instance, in the case of Newton law, they are "force" and "acceleration". On the other hand, many natural, engineering and social phenomena evolve according to non-linear principles, which have been ignored for many years. Linearity is valid at very small time durations and space intervals and any non-linear trace can be considered as a succession of linear parts each during a finite length. Derivation of many non-linear equations and differential equations can be achieved under a set of assumptions, and hence, they are approximations to real situations in natural. As a result of such assumptions, even a slight difference, especially in the initial or boundary conditions, may lead to different output patterns. Hence, although the equations are completely deterministic, their solutions may appear in the form of chaotic behaviors along attractive patterns. Hence, one cannot predict where the next step will be on this attractive orbit, but the completeness of the orbit can be determined.

In order to shed light onto the scientific developments and evolutions in the future the following points can be stated descriptively and briefly.

- Scientific findings are not divine laws. They were not and some others still are not known during certain time periods, but rational human thinking and philosophical principles provide almost certainty for some of the unknowns. Similar to the expansion of the universe and movement of continent (plate tectonics), all the natural phenomena and those that are viewed as deterministic, are in continuous change, and therefore, continuous scientific and engineering works are necessary for their explanation, description and design for the service of humans through, scientific and technological innovations,
- One should not forget that any scientific document is prone to criticism, and hence, continuous development and evolution are in steady existence and human rational thinking must try to capture these facts with improvements from time to time during the scientific evolution. For this purpose, all the time researchers must run after the falsifiability of the scientific principles for the search and arrival at better level including engineering aspects,
- It is necessary to take lessons from the science and engineering history and use such information in future research directions. One of the common mistakes is to think that known information up to present time in any topic covers even the unknowns. This is one of the pitfalls for many researchers in the world, and hence, they cannot make original innovative studies but traditional and imitative works. This principle is correct for geographical and engineering inventions, but scientific inventions are available in the uncertainty world. Today, even in the medical area only 30–40 % is known and there are many research areas in the unknowns' world,

- Intensive researches must be directed by uncertainty methods at the micro scales (particle scale). The author believes that there will always be tremendous research aspects until the end of the universe. Accordingly, engineering activities will also develop in an accumulative and more productive manner. Those who do not take the uncertainty side of scientific inventions cannot make even marginal innovations and they may use science for their personal or ideological purposes, which make the society stagnant in science and technology aspects with traditional, imitative and copy-paste engineering activities. They may think that the present level of scientific knowledge and information are necessary for relevance of any problem. Real scientists can be outraged by such scientism advocators due to their complete faith and dogmatic belief in the present scientific principles,
- One must not forget himself/herself among the scientific activities, because if one does not know himself/herself that the scientific activities cannot be objective and beneficial.

Science and its today's products as technology and engineering applications must develop within peace, equilibrium, regularity, right and justice domains. Otherwise, science cannot develop in a society that does not have these characteristics and they remain the same for the scientist as well as engineers. Of course, it is not possible to annotate all of these characteristics to science only.

7.7 Expert View

An engineer can promote his/her knowledge level by education through frequent works and meaningful rational deductions. Expert views can be defined as the ability of an individual to activate his/her knowledge by specific and personal experience, interrogation, critics and rational thinking for reaching useful and beneficial propositions. If s/he does these frequently by repetition without any criticism, then s/he will be abiding automatically by memorization, artificiality, frozen and non-generative knowledge, which cannot be even labeled as a technician work. Technicians are mediators between the client and engineers, who should also have dynamic knowledge and information. Any technician with dynamic thinking methodology can reach even to the level of knowledge and abilities more than an engineer. Even though s/he is a technician, by training and educating himself/herself s/he may become expert in his/her area, which counts more than a frozen certificate of engineering in practical life. Unfortunately, in some societies expert view certificates and diplomas are more attractive than actual, active and productive expert views and experts. However, in developed societies more than a certificate, the demand is for experts and they can earn more than a classical engineer without expert training after graduation.

For an engineer to increase his/her expert view during the education period s/he must give weight to design aspects of engineering. After the graduation s/he must attach significance to critical thinking and linguistic knowledge more than mathematical formulations, so as to generate first linguistic/verbal solutions through design aspects. Expert view cannot be owned at equal level by each individual. Each expert view has some deviations from others even at slight differences. Such differences indicate that the experts do not have memorized knowledge and information. In these differences are hidden new directions, researches, impressions, discoveries, innovations and possible scientific, technological and engineering research and development clues. For expert view gaining books, lectures and handbooks are not enough. It is necessary to have friction with other experts and with those who are more enlightened on the same or similar aspects.

The most important key in gaining expert view is to criticize existing information and knowledge with their appreciation from different angles in the mind, in addition to the concept, design and related propositions succeeded by logical deductions. During the last four to five decades, engineering education was based on symbolic logic through mathematical expressions and formulations, ready algorithms and the textbooks were counted among the most important means of learning. However, recently, human thinking, brain structure, gens and logical principles with verbal knowledge and information started to gain increasing acceleration that lead to design with approximate (uncertain, fuzzy) methodologies. Among such methodologies artificial intelligence approaches, genetic algorithms, fuzzy logic expert systems and alike play significant role in arriving to the best, optimum, short time, low cost, easy and fast solutions by use of computers [2, 14, 15]. At the foundation of all these methodologies lie verbal information, logic and inferences.

Today in many government institutions and private companies there are expert view empowered engineers, who can reach fast and satisfactory solutions with their experiences. In some societies, unfortunately, rather than being educated and expert in the subject, titles play important role, and hence, experts without title are not respected highly. This is not the case in developed countries, where practical and expert views are sought more than classical academic training. Title is never a power, but "knowledge is power" with "expert views."

There may be a consensus that everybody agree and use the same knowledge, but this does not mean that they all believe in it equally, hence again a philosophical issue emerges. Know-how is one of the modern terminologies in our day and it does not include belief ingredient but practically applicable knowledge. Quite frequently engineers resort to heuristic knowledge and rules of thumb methodologies because there are subjectivity and uncertainty in the knowledge, and hence, such knowledge types can be criticized only on linguistic basis, where philosophy of engineering is needed. Oral engineering is gaining significance by time, because dogmatic and mechanical engineering trainings do not give spiritual feelings, which are also among the triggers for ambitious researches. Unfortunately, today engineering trainings and education are based on automation, which is another key area where simply viewing knowledge as belief is misleading. Furthermore, expert systems and views gained tremendous significance in today's engineering developments, which all need linguistic foundations to start the problem solution. As long as expert views are in question and they differ from each other to a certain extent again the principles of philosophy seeps into the engineering thoughts. In expert systems, routine procedures encode a huge amount of knowledge in corporate memory, and data mining. There are many areas where no one knows the procedures; they are all performed by computers, or systems, or they are distributed over large numbers of people. No one can criticize especially input and output data, because they are sucked down from the available sites. Inference is automated, and therefore, there is no room for debate or discussion. These are among the dead ends in front of engineering because there is no room for debate or discussion and philosophy of engineering is missing. This does not mean that philosophy of engineering is non-existent completely from individual engineers, but in the engineering education system. Automation leads to a huge set of knowledge-based procedures, where input is converted to output and nobody can interfere with the system or interpretation, which is frequently what happens nowadays. In addition to epistemology philosophy of mind (for engineers engineering) is necessary.

7.7.1 Engineering Expert View

Engineering education at graduation time does not provide an expert view knowledge and information. Practical training periods during the engineering education help to gain effective abilities by transferring theoretical information into practical applications. At this stage, the candidate engineer may question whether the level of information is at the stage of "knowing" or "knowledge". S/he may then decide whether s/he completed engineering education as a stage of "knowing" or "knowledge". If the graduation is at the level of "knowing" then s/he starts to learn the things that s/he did not know before. S/he continues the career by letting others to know only. However, if the education is terminated by "knowledge" then s/he will continue the career by questioning every piece of information in order to reach at a higher level of knowledge. S/he will then not only teach but also transfer the questioning abilities to others so as to reach ripeness of the knowledge.

If an engineer has taken education without any question, criticism and doubt (skepticism) about the knowledge content, s/he may start to do so after the graduation in career life. Hence, the engineer starts to leave "knowing" stage and wants to enter and empower himself/herself with "knowledge" and "know how". S/he starts to be acquaintant with philosophical aspects of the knowledge process by questioning the reasons. Hence, approximate reasoning helps to promote his/her expert views. Apart from the experience, there is also experimentation, which leads to real or empirical knowledge. In engineering, empirical relationships are proposed either with rational thinking (see rational matrix in Chap. 5) or data processing after the necessary measurements. The life is full of experience and experiments. The sense of these by each individual is quite different with uncertainty ingredients; for those who will not judge by reasoning the consequences, they will be robotic similar to automation without any new knowledge generation; however, who cares for reasoning will gain new insights, and hence, dynamize his/her information content with beneficial consequences. In fact, for any human real life process starts from birth until death with knowledge software. However, as the age progresses, s/he can revive his/her knowledge by reasoning, and hence, static knowledge gain dynamism. Whoever is after revitalization s/he will appreciate that knowledge storage is not dependent only on dynamic mind experiments but its effect appears on each organ affecting his/her happiness and mind peace.

Recently, the most frequently spelled out words are "expertise", "expert", "expert systems", "expert views", and alike. An engineer may gain expertise knowingly or without knowing if the problems are solved by rational reasoning at every stage of the solution process. In this manner engineer gains extra abilities, which may trigger further problem solutions design proposals and imaginations. If knowledge atomization is based on rational foundations then its rudder will be towards the common benefit of humanity. Otherwise, the rudder may be locked and the direction may be harmful because the knowledge ship may rack on a coast with damages not only on those present on the coastal area but also on itself. One can remember that about 4–5 decades ago, expert views were not counted as scientific aspects, but today they have precious values. The reason for this is that the knowledge that are static lead to similar static solutions, which do not give comfort to human souls and physical appearances, and therefore, differences in solutions are sought through linguistic information and suggestions because engineering algorithms, formulations and equations lead to deterministic consequences.

In order to suggest expert views linguistic information should be debatable. Recently, this is referred to as "know-how", which means to know how the linguistic solutions work in the solution of any problem. Linguistic explanations change from engineer to others to a fuzzy extends, and hence, multiple overlapping views emerge, where main clues may lie for the best solution. In a classical and non-interrogative education system, standard and the same views are common and unique solutions are dominant, which do not have any science philosophical basis but hidden logical statements that are not well-known by engineers. This leads to a single color rather than a mosaic of colors for different opinion generations. In the latter case differences in opinion between various experts help to reach to a common, best, optimum and the most convenient solution after mutual dialogs.

Engineer must try to develop not only his/her physical and physiological body abilities but through research activities in the service of the society with practically applicable end products. Among the most important abilities are fine arts, sports, spiritual activities, views about social events and their discussion with others, optimum management of time and management practices. All these activities assist an engineer to reach at rational conclusions with creative ideas through the use of experience, experiments and trial and error scenarios. Additionally, expert views give way to team works with transparency, clarity and generality. Hence, intelligent collective work provides dynamism for creative ideas and different solution generations with the best selection among them. The fundamental of intelligence is the dynamism of the perceptions through all organs, which also work in a collective manner. Such activities can be molded into useful deductions after filtrations through verbal logic and its aftermath as symbolic logic in engineering. Contribution of all concerned individuals collectively on an issue paves way after critical debates to common intelligent products.

Expert systems may evolve through either supervised or unsupervised trainings and likewise engineering mind has also the same training branches for human intellectual mind exploration. Different engineers may adapt various conceptualizations about an event for predicting and the unfolding of its generation mechanism dynamics, which cannot be achieved by absolute control or prediction. Eager engineers bring forward a set of acceptable solutions and then choose the most convenient one on the bases of economics, simplicity, aesthetics and optimality. Such a goal can be achieved by considering the logical interdependence among various fragments of the whole.

Who is not empowered with expert view cannot make decisions in front of a set of alternative solutions. Decision-making is among the soft skill abilities of an engineer, but it may not be triggered if the engineering candidates have not been alerted about its significance. Unfortunately, it is not thought during the engineering education and after the graduation at work they depend on given prescriptional and ready classical solutions without alternatives. Accordingly, without decision making capability they stick to "do blindly as thought" and in the long run "continue current approach" as the only valid options. For reaching to effective and plausible decisions as conclusions one should ask to himself/herself "Am I thinking and progressing efficiently? Is there any better, shorter, cheaper, faster and simple way to complete the task?". Such questions will help young engineer to train himself/herself in the right direction and in this way s/he will get more excited about alternative problem solution and firm decision making among the set of alternatives. Decisiveness makes engineer more eager to achieve, and hence s/he will taste the happiness in the career.

7.8 Future Reflections from the Past

Successful application of the classical control systems have been appreciated in industrial and engineering solutions. However, there remains still uncertainties to a certain extend that cannot be modeled by the classical approaches, and therefore, uncertainty assessment methodologies are necessary. On this regard for many years probabilistic, statistical and stochastic approaches and methods have been exploited to the farthest extend with effective impacts of linguistic concepts. Due to the lack of verbal information content, at times main knowledge could not be appended to the whole system. On the other hand their inclusion in the system brings further dimensions and additions which may be unmanageable to solve with certainty. In this respect, for the last three decades and especially during the last decade increasingly the use of fuzzy systems approach has penetrated many branches of sciences and preferably technology and engineering. Modern industries and technologies require increased flexibility, which results

Fig. 7.1 Design from the twelth century (Abou-l Iz Al-Jazari, 1136–1206)

in highly nonlinear system behaviors that are known only partially with remaining uncertainties in the main system. These uncertainties may be negligibly small but their collective and especially cumulative effects give rise to complex solutions that cannot be determined uniquely. The classical control systems whatever their advancement levels are, they can only partly be able to satisfy the demands. However, the fuzzy logic and system analysis can give at this junction help to satisfy the demand through qualitative operator and design knowledge for implementations. In the past and still currently in a decreasing manner the fuzzy concepts have led to misunderstandings and controversial issues between many researchers.

Prior to differential and integral calculus clever enough human beings have been trying to work in an artistic manner like a craftsman to make many devices by using their common sense, and experiences to design water wheels in order to rise water from a lower elevation to higher positions, wind mills, water pumps as shown in Fig. 7.1. This figure is due to Muslim workmen who made during the twelfth century automatically working water machines.

It is obvious from this figure that even at the twelfth century the cylinder, piston and valve pieces are put collectively in a system for water pumping. Of course, such designs have always existed in the literature before the systematic, mathematical and differential thinking. This proves the point that human is able to design their works not depending on higher mathematics or education but by their ability to think towards a purpose. Since thinking is a verbal (linguistic) process, it is necessary that the preliminary concepts and imaginations for any purpose will be in words not in equations as we are unfortunately very fond even addicted to on these days. Genuine thinking and imaginations may lead to conceptual and physical design concepts and virtually to generation of opinions towards this end.

It is hardly possible to find philosophical ingredients in technological aspects except recently. Engineering and philosophy are regarded as almost completely separate from each other. However, most often one can hear about the philosophy of science, which is more established than the philosophy entrance into the technology and engineering. The most frequently met philosophical subjects since 1980s appear in ethics, which is now in the curriculum of many universities, not directly coupled with philosophy, but at least due to its nature philosophical aspects are implied in explanations. It is also interesting to think about the distinction, if any, between the philosophy of technology and engineering. In order to answer to this question one should first know what is the meaning of technology and engineering. The view taken in this book is that engineering and technology are not distinct from each other. One can also think about creative works of engineers towards technological innovations even though there is a slight improvement in the existing instrumentations and gadgets. During the generative ideas without deep understanding of the philosophy they do their works, but even so there are faint philosophical ingredients in the works. This can be regarded as traditional or common philosophy (systematic or haphazard thinking) but not real philosophical reflections. It is evident recently that the engineers need more philosophical demise than ever before to suggest innovative developments or critical assessment at the existing information and knowledge level.

References

1. Luce RD, Krantz DH, Suppes P, Tversky A (1990) Foundations of Measurement, vol III. Academic Press, Inc, New York
2. Şen, Z (2010a) Fuzzy Logic and Hydrological Modeling. Taylor and FrancisGroup, CRC Press, Boca Raton 340
3. Şen, Z. (2010b) Rapid visual earthquake hazard evaluation of existingbuildings by fuzzy. Expert Systems with Applications. 37:5653–5660
4. Stevens SS (1946) On the theory of scales of measurement. Science 103:677–680
5. Stevens SS (1951) Mathematics, measurement, and psychophysics. In: Stevens SS (ed) Handbook of experimental psychology. John Wiley, New York
6. Luce RD (1959) On the possible psychophysical laws. Psychol Rev 66:81–95
7. Zadeh LA., (1965) Fuzzy sets. Information and Control 8: 338–353
8. Shannon CE, Wiener W (1949) The mathematical theory of communication urban. University of Illinois Press, Urbana, p 125
9. Carlton AG, Follin JW Jr (1956) Recent developments in fixed and adaptive filtering. In: AGARDograph 21
10. Kalman RE (1960) A new approach to linear filtering and prediction problems. ASME Trans, J Basic Eng Ser D 82:35–45
11. Kuhn TS (1996) The structure of scientific revolutions, 3rd edn. University of Chicago Press, Chicago
12. Mandelbrot B (1978) The fractal geometry of nature. Freeman, New York
13. Zadeh LA (1978) Fuzzy sets as a basis for a theory of possibility. Fuzzy Sets Syst 1:3–28
14. Şen, Z (2004a) Yapay Sinir Aglari (Artificial Neural Networks) Su VakfıYayinlari (Turkish Water Foundation Publications), pp 312 (in Turkish)
15. Şen, Z (2004b) Genetik Algoritmalar (Genetic Algorithms) Su VakfıYayinlari (Turkish Water Foundation Publications), pp 282 (in Turkish)

Index

Printed in the United States
By Bookmasters